Sustainable Supply Chain Management

For my parents

Sustainable Supply Chain Management

Joëlle Morana

Series Editor
Jean-Paul Bourrières

iSTE WILEY

First published 2013 in Great Britain and the United States by ISTE Ltd and John Wiley & Sons, Inc.

ISTE Ltd
27-37 St George's Road
London SW19 4EU
UK

www.iste.co.uk

John Wiley & Sons, Inc.
111 River Street
Hoboken, NJ 07030
USA

www.wiley.com

Library of Congress Control Number: 2013934673

British Library Cataloguing-in-Publication Data
A CIP record for this book is available from the British Library
ISBN: 978-1-84821-526-9

Table of Contents

Acknowledgements

> "Be like the bird who, pausing in her flight awhile on boughs too slight, feels them give way beneath her, and yet sings, knowing she hath wings."
>
> Victor Hugo (1802-1885)

This book would never have seen the light of day without the support of a number of people, to whom I would like to express my heartfelt thanks.

Throughout the writing of this book, a significant part has been played by people who, while they were not center-stage, have enriched the manuscript, with sound bibliographical advice, careful reading and unwavering friendship. In particular, I would like to thank the following people for their time in reading this manuscript and for their valuable advice:

M. Maurice Bernadet, Professor Emeritus at the Laboratoire d'Economie des Transports (LET – Transport Economics Laboratory) at Université Lumière Lyon 2, Lyon, France;

Mme Michèle Combe, the librarian at IAE, Aix-en-Provence, France;

M. Sergio Fernando Fortun Auad, a doctoral candidate at IAE, Aix-en-Provence. France;

Mme Martine Sefsaf, the archivist at the LET at University Lumière Lyon 2, Lyon, France.

I am also grateful to a number of members of the LET for their moral support. The people in question know who they are.

Introduction

I.1. Introduction

It is widely recognized today that supply chain management has become a matter of strategic importance for any company, in the primary, secondary and/or tertiary sectors. For reasons of external pressures, globalization and competition, supply chain management optimizes the processes and reduces the cycles of production and delivery. Put simply, supply chain management enables traders to adapt to the numerous changing factors in the market. Daily, we hear the examples cited of the economic, environmental and social/societal constraints which mean every company – and therefore every supply chain – must always be at the cutting edge of innovations – technological and/or organizational. Faced with this situation, it is certain that in the near future, all companies will have to exercise strategic logistics in general, but also: Sustainable Supply Chain Management

I.2. Historical background on how supply chain management has become strategic and omnipresent

Historically, companies only began to exercise logistical strategy in the 1950s. However, as Colin [COL 96] points out, it was through the construction of the French Navy in the 18th Century that "the main concepts which are now at the root of the approach to logistics practiced today" emerged and took shape. Of these concepts, we can cite anticipation (branding of those trees intended for the purposes of shipbuilding), reactivity (the range of seafaring people, including merchant sailors and fishermen), continuity and reliability (building of canals to guarantee constant supplies) and standardization (limitation of the series of ships). This standardization improved time management, cost management, availability of parts because of interchangeability, flexibility of the fleet, which gave rise to an aptitude for prediction and, in fact, enhanced the quality of the vessels built. Even today, these concepts are the fundamental elements in all companies' logistical approach.

Sticking with the historical context, if we look at its development in the world of business, Colin [COL 02] gives an overview, dividing it into two main periods. Firstly, until the 1960s, mass production was in evidence. We speak of a "gravity model", characterized by shortage and with which standard goods with long series and short cycles are the norm. The level of service is therefore poorer than poor, and supply chain logistics are not the panacea. At this level, one would tend to speak of a supply chain with push-flow factors, where the necessity for production governs the whole of the supply chain. The information flows in the same direction as the physical flow: from upstream to downstream (producers to consumers). Then, after the 1960s, competition led to personalization, and to "made-to-measure", bespoke products. This saw the emergence of an alternative model: the "market-regulated model". The switch from the first to the second model entailed enormous structural and organizational change: "moving from managing demand on the basis of supply to managing supply on the basis of demand required a 180° reversal of priorities, and completely overturned the structures of the firm"[COL 02]. Gradually, logistics went from being a "technical" vision (in the 1970s–1980s), where we seek to optimize the tasks, to being a "service" vision (in the 1980s–1990s), aiming for better delivery times, for instance. At this stage, logistical regulation will be identified by a "pull-flow" system, whereby the production process, and therefore the financial flow, is begun on the basis of the demand. The information flows from downstream to upstream (consumer to manufacturer), giving the order for manufacture.

Since this period, logistics has become transactional and relational in nature, both from an intra- and inter-organizational point of view. In that sense, it becomes a combination of three "primary" flows: flow of goods, flow of information and financial flow [AUR 97a]. In doing so, a fourth flow is also set alongside these three flows: the flow of people who interact in the design and management of the overall system [HES 73; TIX 83]. Mesnard and Dupont [MES 99], from the consultancy firm AT Kearney, argue for the inclusion of a flow of intelligence which, beyond the traditional flows (goods, information, finance, people), facilitates maximum exploitation of all forms of information. Thus, for these authors:

– logistics should instill a reactive capability, meaning the company is able to identify and satisfy unforeseen demands. On the one hand, this requires the mobilization of significant material and immaterial investment [GER 98; CHE 98]; on the other hand, it requires flexibility in order to react to unexpected circumstances [PON 93; REI 97; TAR 99; EST 99];

– agile logistics refers to the ability to swiftly redistribute the available resources from one end of the supply chain to the other [CHR 99];

– in order to be effective in the long term, companies need to seek to maximize the synergy between the various logistical chains in which they are involved. Hence, a company needs to be efficient, firstly, in systematically eliminating all forms of waste and, secondly, in optimizing the monitoring of the overhead costs, the logistical processes, the resources committed and the costs external to the environment.

Yet in order to speak of a company's logistical strategy, we need to discuss the concept of *supply chain management*. In general, while so-called "operational logistics" is appreciable from the point of view of logistical strategy, and conversely, so-called "strategic logistics" highlights all the advantages to strategic logistical practices, the term" supply chain management" now supplants these two notions, encapsulating both of them.

Supply chain management (SCM) has witnessed particular development since the seminal work published by Christopher [CHR 92], *Logistics and Supply Chain Management*. At the time, the author defined supply chain management as:

> "a network of organizations that are involved, through upstream and downstream linkages, in the different processes and activities that produce value in the form of products and services in the hands of the ultimate customer."

Many more definitions would subsequently be put forward, but all with the same direction – i.e. the quest for cooperation and collaboration to ensure and improve the performance of each company included in a supply chain.

However, in reviewing the existing literature today, we see two main definitions emerge: that of Mentzer *et al.* [MEN 01] and that advanced by the *Council of Supply Chain Management Professionals* (CSCMP). These two definitions are presented in Chapter 1.

I.3. The emergence of sustainable supply chain management

While the contribution of supply chain management is becoming ever more significant, it does not stop at the economic aspect. It is also to be found in ecosystem preservation [PAC 99; PAC 00]. The same is true of social and societal recognition of the actors who make up the supply chain. It is therefore becoming appropriate to speak of *sustainable supply chain management*, which is akin to *sustainable development*.

The issue thus arises of *sustainable SCM* (SuSCM). According to Seuring and Müller [SEU 08], it is possible to define:

> "[…] sustainable supply chain management as the management of material, information and capital flows as well as cooperation among companies along the supply chain while taking goals from all three dimensions of

> sustainable development, i.e. economic, environmental and social, into account which are derived from customer and stakeholders requirements."

Although it is rather general, this definition has the advantage of demonstrating the link between supply chain management and sustainable development. This being the case, a question arises about this definition. Indeed, in stakeholder theory, the customer is included as part of the group of stakeholders [HIL 92; DON 95]. Here, however, these two authors appear to consider the clients as being separate from the stakeholder. Thus, in our view, it would have been more appropriate to speak of "the customer and the other stakeholders" or to leave "the stakeholders' expectations". On the same point, a great many authors in the field of logistics speak of the *end client* or the *final customer*. Thus, a distinction is to be drawn between the customer at the end of the chain and the customer – in the general sense – who is present throughout the length of the supply chain, within and outside of every company. This element should therefore have been taken into consideration in the definition put forward.

According to Carter and Rogers [CAR 08], SuSCM is defined as:

> "[...] the strategic, transparent integration and achievement of an organization's social, environmental, and economic goals in the systemic coordination of key interorganizational business processes for improving the long-term economic performance of the individual company and its supply chains."

For our part, we prefer the following definition:

> "Sustainable Supply Chain Management can be understood as the management of the flows of materials, information, capital, people and intelligence with an

> economic, environmental and social/societal[1] purpose. As a strategic management approach, it is to be found in the quite deliberate set of intra- and inter-organizational connections, with a view to the long-term performance of each company and of its supply chain."

If the practice of SuSCM is becoming an increasingly important theme for discussion, it is because there is pressure from numerous sources for it to be implemented: the requirements of various stakeholders, governmental pressure (decrees, laws, norms/standards, etc.), environmental pressure (pollution, exhaustion of fossil fuels, etc.) and social/societal pressure (reputation/image, protection, etc.) [SEU 08].

Nevertheless, there are certain barriers to the implementation of SuSCM. Of these, we generally cite significant costs, complexity of coordination and insufficient communication in any supply chain. There is also the fact that all organizations now have to include a sustainable strategic approach when making its decisions. Support mechanisms are in place to help with this – supports based on information and communication technology (ICT), management systems (e.g. the ISO 26000 standard, the ISO 14000 standard, the SA 8000 standard, etc.), training, etc. Similarly, there are sanction mechanisms such as those stemming from the practice of traceability (e.g. the "Polluter Pays Principle") or indeed decrees such as directives 2002/95/EC and 2002/96/EC on Waste Electrical and Electronic Equipment (WEEE).

---

1 In the context of sustainable development, authors often place the notions of "social" and "societal" at the same level. However, the term "social" should be used in preference when speaking about actions directed at the company's employees. The term "societal", on the other hand, should be used to speak of actions directed at all the stakeholders. (See Chapter 3,"Social/Societal Supply Chain Management", for a discussion of the social and societal impacts). In fact, in this book, we use these two terms in juxtaposition.

In any case, it is becoming increasingly advantageous to practice SuSCM. In order to do so, we need to draw on the work carried out in the field of logistics, or in other domains of management sciences (human resources, quality assurance, strategy, management control, etc.), taking from this work the methods to facilitate the exercise of a SuSCM strategy. With this in mind, this book is divided as follows:

– Chapter 1 will deal with the economic aspect of SuSCM, which is particularly appropriate to the works associated with supply chain management;

– Chapter 2 will look at the environmental aspect of SuSCM, which corresponds closely with Green SCM (GrSCM);

– Chapter 3 will investigate the social/societal aspect of SuSCM, which finds its place in a Social/Societal SCM (2SoSCM), for which a particular arrangement needs to be set out.

Yet it must be noted that, although the performance and effectiveness of the approach can be quantified on the basis of each of these aspects, it seems more useful to establish a "final" consolidation – a *sustainable supply chain management balanced scorecard* – which uses appropriate *Key Performance Indicators* to support the measurement of these aspects. This balanced scorecard approach will be the subject of Chapter 4.

Thus, the aim of this book is to give a detailed discussion of each of the aspects – economic, environmental and social/societal – of SuSCM. Considering each of these aspects separately, the primary objective is to facilitate the implementation of the elements of each of these approaches. A secondary objective is to provide the reader with as many "strategic grids" as possible to enable him to carry out audits and diagnostics for each component. Hence, this book approaches SuSCM from an "upstream" standpoint, aiming to answer the general question: What are the elements that I need to take into consideration in order to implement *sustainable supply chain management*?

Figure I.1 illustrates this configuration.

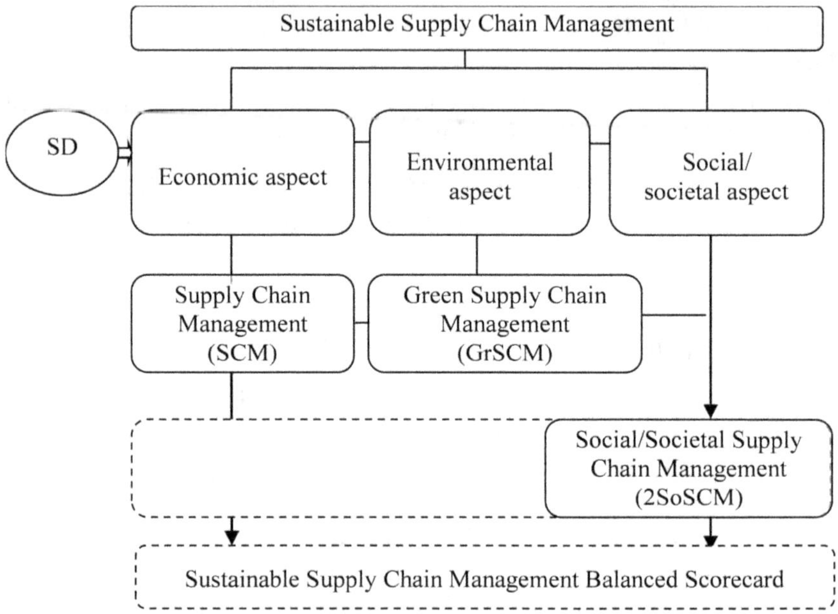

Figure I.1. *The components of sustainable supply chain management (SD = sustainable development)*

Chapter 1

The Economic Aspect of Sustainable Supply Chain Management

1.1. Introduction

Globalization and increasingly fierce competition, among the most frequently cited factors, have led managers to begin to examine the question of more efficient logistics. Thus, from the 1980s onwards, we see a move away from logistics with an "operations-oriented" status (we speak of "logistical strategy") towards one with a "strategy-oriented" status (we speak of "strategic logistics"). Yet in modern-day language, the term "supply chain management" is predominantly used.

A review of the literature advocating supply chain management illustrates the important role that it plays in companies' decision-making and in the thinking of researchers in this field. By way of example, we can cite the many works which give an overview of the reviews and/or conferences which highlight the prolific recourse to supply chain management and to the aspects which make it up: the analysis of 20 years of the *Journal of Business Logistics* [MIY 99], that of the first 10 and the first 16 years of the journal *Logistique et Management* [MOR 04a; MOR 05; MOR 10][1], the bibliometric

1 The analysis of the first 16 years of the journal *Logistique et Management* [MOR 10] highlights eight themes which seem to dominate the thinking of logistical

review of the first four and first five *Rencontres Internationales de Recherche en Logistique* (International Conferences on Research in Logistics) [BOI 04; BOI 06] or indeed the analysis of the 124 articles and conferences in *Strategy*[2] referring to logistics and supply chain management [CHE 06]. From these analyses, we see the transversality of supply chain management in all the fields relating to management of an organization.

Thus, the use of supply chain management has become an absolute necessity for companies. It can in fact be classed as a *raison d'être* for any organization that is part of a supply chain. To overlook it is a risk that no company is willing to take.

1.1.1. *The emergence of supply chain management*

Although the term "supply chain management" was introduced by Christopher [CHR 92], there are strong parallels to be drawn with the concept of "*logistique*" (strategic logistics) developed by a French school of thought [MOR 02]. The difference, if any, lies in the fact that supply chain management includes a "manifest and mutual" quest for alliance and cooperation, and the risks and rewards are shared between the participants [MEN 01]. According to Harland [HAR 96], we can distinguish supply chain management by the number of firms involved in the logistical project [JOS 83]. For Harland [HAR 96], (p. 64), supply chain management – in its purest form – is to be considered as "the management of a network of interconnected businesses involved in the ultimate provision of product and service packages required by end customers".

---

researchers. These are, successively: (1) supply chain management, with all the elements relating to intra- and inter-organizational structure; (2) logistics providers/services; (3) logistical practices where we find discussions about upstream, production and downstream logistics; (4) management of information and communication technology; (5) transport, which includes problems relating to zoning laws, for instance; (6) performance evaluation; (7) human resource management; and (8) sustainable development, with particular focus on ecology/the environment.

2 *Strategic Management Journal*, *Academy of Management Journal*, *Management Science*, *Organization Science*, *Administrative Science Quarterly* and the conference of the Academy of Management.

By a review of the literature relating to logistics, we are able to extract the definitions attached to supply chain management. Of the various definitions advanced in this body of literature, we shall choose three which appear to reflect the salient elements.

Historically, the definition put forward by Christopher [CHR 92] has served as a reference point:

> Supply chain management is a network of organizations that are involved through upstream and downstream linkages in the different processes and activities that produce value in the form of products and services in the hands of the ultimate customer.

However, the definitions which are most widely used appear to be those stemming from the work of Mentzer *et al.* [MEN 01] and that of the Council of Supply Chain Management Professionals (CSCMP).

Mentzer *et al.*'s definition [MEN 01] is presented thus:

> Supply chain management is the systematic, strategic coordination of the traditional business functions and the tactics across these business functions within a particular company and across businesses within the supply chain, for the purposes of improving the long-term performance of the individual companies and the supply chain as a whole.

The CSCMP's definition is the result of an in-depth study put to the members of the council (sample = 744 usable responses from the 6,422 members questioned [GIB 05]):

> Supply Chain Management encompasses the planning and management of all activities involved in sourcing and procurement, conversion, and all Logistics Management activities. <u>Importantly</u>, it also includes coordination and collaboration with channel partners, which can be suppliers, intermediaries, third-party service providers, and customers. In essence, Supply Chain Management

> integrates supply and demand management within and across companies.

This definition is supplemented by the following statement on the Council of Supply Chain Management Professionals (CSCMP) website (www.cscmp.org):

> Supply Chain Management is an integrating function with primary responsibility for linking major business functions and business processes within and across companies into a cohesive and high-performing business model. It includes all of the logistics management activities noted above, as well as manufacturing operations, and it drives coordination of processes and activities with and across marketing, sales, product design, finance and information technology.

Council of Supply Chain Management Professionals (CSCMP)

On 15 July 2004, the executive committee of the Council of Logistics Management voted to change the name of the organization. On 1 January 2005, it became the Council of Supply Chain Management Professionals.

This is a not-for-profit professional organization whose goal is to develop and diffuse as much information as possible relating to supply chain management.

In 2012, the CSCMP had over 8,500 members – professionals from 65 countries, covering six of the seven continents.

1.1.2. *The aspects of supply chain management*

Many schools and many authors define the component aspects of supply chain management. Here, we present three of these schools/works, which we deem to be influential in this discussion: Michigan State University's "supply chain structure 2000" [GRL 95; BOW 99]; Ohio State University [COO 97; LAM 98; LAM 05] and the publications by Mentzer *et al.*'s working group [MEN 01] followed by the research of Min and Mentzer [MIN 04].

In the view of Michigan State University [GRL 95; BOW 99], three "contexts" are associated with supply chain management:

– the "operational" context recommends the integration of the stakeholders in a joint project. The aim is to maximize the forces of standardization between the upstream, midstream and downstream actors in the supply chain;

– the "planning and control" context strives for compatibility of the ICT elements. By means of optimal technology, planning and control, the goal is to enable the various stakeholders to be aware of the status of the systems and to make the various resources that are present consistent;

– the "behavioral" context highlights the coordination between the actors, defining the roles of each one and sharing the risks and rewards.

In the work published by Ohio State University [COO 97; LAM 98], there are three elements which make up supply chain management:

– the "relational structure of the supply chain network": the aim here is to identify the number and critical positioning of the actors involved in the chain;

– the "managerial processes of the supply chain": the goal is to identify and analyze the processes for each activity;

– the "component management": this is divided into:

- a "physical and technical component" which includes: (1) planning and control methods; (2) the work patterns/structure of the activity; (3) the structure of the organization; (4) communication and information flow of the technical structure; and (5) the product flow of the technical structure,

- a "managerial and behavioral component", which includes: (1) management methods; (2) the power and leadership structure; (3) the structure of risks and rewards; and (4) the culture and attitude.

Finally, in the third group of work, we find the writings of Mentzer *et al.* [MEN 01]. These authors structure supply chain management with two additional elements:

– "Supply Chain Orientation" (SCO): "the recognition by an organization of the systemic, strategic implications of the tactical activities involved in managing the various flows in a supply chain"; and

– "Supply Chain Management" (SCM): "Supply chain management is the systematic, strategic coordination of the traditional business functions and the tactics across these business functions within a particular company and across businesses within the supply chain, for the purposes of improving the long-term performance of the individual companies and the supply chain as a whole" (p. 18).

SCM can come into play when SCO is applied. On the basis of these two definitions, Min and Mentzer [MIN 04] put forward a list of items that can be used to evaluate SCM (in the broadest sense of the term) in the different dimensions which make it up. Thus:

– the first scale, SCO, includes six factors: credibility, benevolence, commitment, norms, compatibility and top management support;

– the second scale, SCM, includes seven factors: agreed vision and goals, information-sharing, risk and reward sharing, cooperation, process integration, long-term relationship and agreed supply chain leadership;

– the third scale, PERF (performance), includes five factors: availability, product and service offerings, timeliness, profitability and growth.

1.1.3. *Overview of the dimensions of supply chain management: proposition*

Finally, Tables 1.1, 1.2 and 1.3 give an overview of the thinking of the three schools/works discussed above. This layout constitutes an analytical grid which enables us to easily see the common threads running through these models. However, it should be pointed out that these tables do not highlight the item "transport", which has not been widely discussed in the body of literature on logistics, but whose importance in inter-organizational linkages is stressed by Cheng and Grimm [CHE 06], for instance.

Michigan State University: GRLT (1995); [BOW 99]
Operational context
1. Customer integration
- segmentation of the customers
- relevance of each demand
- response to each demand
- flexibility in unforeseen circumstances
2. Internal integration
- union of departmental functions
- standard, simple and consistent procedures
- structural adaptation
3. Integration of the goods and service provider
- on a strategic, financial, operational and managerial level
Planning and control context
4. Technology and planning
- management of information and of internal communication
- linkage between actors
- estimated budget/planning collaboration
5. Measurement
- functional evaluation
- evaluation methodologies
- metrics
- financial impact

Table 1.1. *The elements of supply chain management according to Michigan State University (source: adapted from GRLT (1995) and [BOW 99]: reprinted from [MOR 07])*

Behavioral context
6. Relational integration - clearly defined roles for everyone involved - conduct standard - information-sharing - risk- and reward-sharing

Table 1.1.(continued) *The elements of supply chain management according to Michigan State University (source: adapted from GRLT (1995) and [BOW 99]: reprinted from [MOR 07])*

Ohio State University: [COO 97; LAM 98; LAM 05]
1. Components of supply chain management - physical and technical (planning and control methods, work patterns/structure of the activity, structure of the organization, communication and information flow of the technical structure, product flow of the technical structure) - managerial and behavioral (management method, power and leadership structure, risk and reward structure, culture and attitude) *2. Value creation*
3. Managerial processes in the supply chain - improving efficacy and decreasing friction - identifying and analyzing the processes of each activity *4. Extended* - strategy - activities
5. Relational structure - number of tiers along the chain - number of suppliers and customers within each of these tiers - identification of the primary and support members - positioning of society on the supply chain *6. Intra- and inter-organizational connections*

Table 1.2. *The elements of supply chain management according to Ohio State University (source: adapted from [COO 97] and [LAM 98; LAM 05]: reprinted from [MOR 07])*

[MEN 01; MIN 04]
1. *Supply chain orientation (SCO)* - credibility: concepts of reliability, recognition of the company, no false claims; - benevolence: "health" of the company, answers to questions, risks and rewards shared; - commitment: cooperation, technical support; - Top Management Support: communication of objectives, long-term contracts, training; - compatibility: shared strategy and culture; - norms: cooperation, value creation.
2. *Supply chain management (SCM)* - agreed vision and goals: standardization of practices, definition of roles; - information-sharing (IS): exchange of predictions, of planning; - risk and reward sharing: financial aid, shared R&D; - cooperation, integration of processes: reports, indicators, quality, ideas box, communication about new products, stock management; - long-term relationships: reduction of complexity, audits; - appropriate leadership of the supply chain: audit, benchmarking.
3. *Performance (PERF)* - availability: stock, backup stock; - supply of products and services: variety, quality in comparison to competitors; - timeliness: just-in-time delivery and information; - profitability: financial return; - growth: revenue, market share.

Table 1.3. *The elements of supply chain management according to Mentzer et al. (source: adapted from [MEN 01] and [MIN 04]: reprinted from [MOR 07])*

In summary, the execution of a SuSCM in its economic aspect is based mainly on intra- and inter-organizational collaboration/coopetition. Thus, the question of whether to use the term "collaboration" which, in general, manifests itself by the practice of work in a spirit of general benefit, or "coopetition", which amalgamates the notions of cooperation and competition, now arises. In actual fact, we have chosen to employ a term which we feel is broader and more flexible: *connection* (granted, a term found more often in the field of electronics!). Some near synonyms are: consistency, cohesion, linkage, link and relation[3], and therefore the term has a parallel in supply chain management. It goes without saying that these intra- and inter-organizational connections must first be recognized as preponderant by the general management team (MT) of each organization involved in a supply chain.

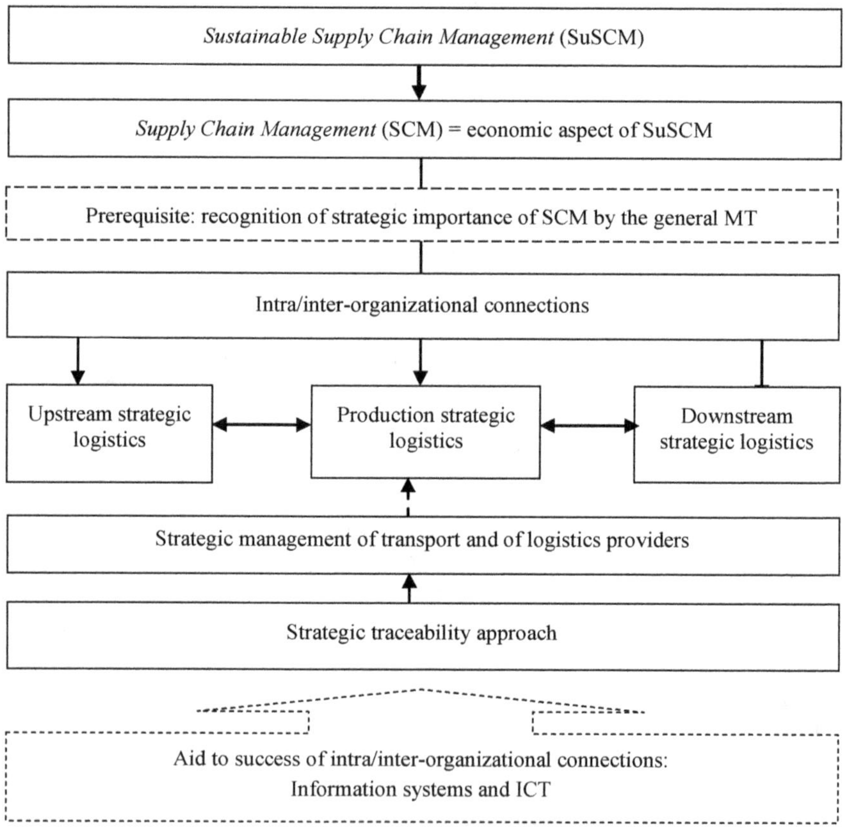

Figure 1.1. *Sustainable supply chain management in its economic aspect*

3 http://thesaurus.com/browse/connection?s=t.

The connections have an impact on the "three logistics" which are often outlined in the description of the logistical strategy of each company involved in the supply chain. Thus, any logistics – be it upstream at the supply and production stage (internal), or downstream at the distribution stage – must be devised and reflected upon strategically. Similarly, in this long-term framework, other elements need to be developed – namely strategic transport management, the role of the logistics providers and the traceability approach. Finally, it must be understood that information and its supports facilitate these connections (see Figure 1.1).

The next part of this chapter is given over to the discussion of each aspect.

1.2. Intra- and inter-organizational connections

Today's society is characterized by new relations between the social actors. Large groups tend no longer to be able to act with total autonomy. In order to manage the distribution circuit of a product, from the extraction of the raw material to the delivery of the finished product, companies need to group together in networks. To facilitate this task, a "destructuration" of all the intra-organizational logistics in an upstream, production/internal or downstream process – all other things being equal – aids decision-making. The aim of the following sections is to put forward tools, matrices and methods which are not widely covered by the literature currently available and which are relevant for these three types of logistics. This section is therefore positioned with a deliberate intention to go further, to put forward elements that can be strategically integrated into the intra- and also inter-organizational connections.

1.2.1. *Strategically upstream or supply logistics*

A properly-controlled beginning provides solid bases to fulfill the objectives defined in advance by the directors as fully as possible. Upstream strategic logistics (also referred to as "supply logistics") guarantees success thanks to the decision-making role of the buyer, who comes to hold an increasing amount of sway in his transactions with a supplier.

Thus, by putting in place new so-called "just in time (JIT)" procedural practices [CAL 98], deciding to outsource [CAL 99], discussing the development of new products [CAL 00a] or indeed constructing common management tools [CAL 00b], we are able to "constrain" the buyer to (accrued) sets of skills in technical and interpersonal relations [CAL 00a; SAU 00].

In the everyday sense of the term, purchasing (of raw materials, processed material, supplies, etc.) is considered to be a supporting activity, in the eyes of Porter [POR 85]. However, in a systemic vision of the company, as is advocated in companies nowadays, the "integrative and strategic" role of the purchasing function becomes critical.

In his strategic musings on the role of the purchasing department, Calvi [CAL 99] raises the question of outsourcing of purchasing activities. In his view, it is helpful to answer two types of question when it is a question of "give-and-take":

– to which types of purchases can we apply this outsourcing?

– which activities in the process of purchasing/supply are involved?

In terms of the types of purchase, there are two categories which could be outsourced. To begin with, we can outsource complex, unusual and non-repetitive purchases, in which case it may be advantageous to draw on outside purchasing skills. Then, we can outsource so-called "tactical" purchases, which are simple and repetitive, and therefore liable to take up a great deal of administrators' time. When deciding to outsource certain operations, there are two elements or hurdles which need to be considered: the risk of loss of confidentiality and the loss of control of the costs relating to the purchase. Hence, as part of a policy of outsourcing, the purchasing activity which can most easily be outsourced is "order-processing".

In general, the buyer tends to play three main roles [CAL 99], ranging from an operational to a strategic role:

– “purchasing management”, which covers all the activities of directing, organizing and monitoring the purchasing service;

– “purchasing marketing”, which prepares for the act of purchase in the normal sense of the term;[4]

– “supplier portfolio management”, where the activities are connected with the act of purchase (order-processing) and the monitoring and development of current relationships with suppliers.

These three activities influence the purchasing/supply process.

1.2.1.1. *The seven stages of the purchasing/supply process*

Seven stages make up the operational process of purchasing and supply:

– first stage: specification of requirements, wherein the buyer:

- brings his knowledge of the market to the table,

- provides information about the prices, the risks involved in the supply, and innovations,

---

4 Barriol [BAR 98] analyzes the history of “purchasing marketing” over a period of 25 years. The author presents three phases:

– the emergence, between 1973 and 1980, where four aspects are highlighted – namely:

- a state of mind: improving the buyer’s position by adopting a more aggressive stance toward the supplier,

- an incentive approach: making the buyer company attractive,

- internal and external marketing,

- the definition of tools shared between marketing and purchasing, such as the concept of *purchasing-mix* (price, product, communication, market);

– a phase of development between 1980 and 1990, where purchasing marketing moved from being a commercial issue to being a purchasing strategy for the company, with the dawn of informational, organizational and communicational tools;

– from 1990 onwards, a conceptual stability which highlights the fact that purchasing marketing combines internal aspects (the company’s needs), external aspects (prior situation of the market), analysis and action (adjustment of the needs and of the market).

- suggests substitute products,

- advocates standardization of the products;

– second, third and fourth stages, which are, respectively, the "search for suppliers", the "choice of suppliers" and "negotiation". Here, the buyer:

- analyzes the offers using a multi-criterion approach,

- selects the best offer,

- negotiates and contractualizes the relationship,

- defines the general framework for the exchange,

- feeds information to the other internal functions involved;

– fifth, sixth and seventh stages, which are "order issuing", "receipt" and "payment". These three stages constitute the "order-processing" phase, which is one of the tasks that can easily be outsourced. For these last three stages of the process, the buyer will first:

- design the order,

- create the order,

- send the order,

- track the order,

- control the supply [order issuing stage].

Then perform:

– recording,

– storage,

– quality control [receipt stage];

and, finally:

– pay the bill,

– examine the invoice,

– deal with any disputes [payment stage].

1.2.1.2. *Purchasing marketing*

Purchasing marketing takes place at the tactical level, and involves four actions:

– analysis of the offer;

– technological monitoring;

– communication with the suppliers;

– viewing and analysis of the purchasing portfolio.

Throughout the process of purchasing/supply, the buyer acts at the first two stages, i.e. "specification of requirements" and "search for suppliers". In order to correctly define the "specification of requirements", the buyer makes contact with and takes account of the opinion of the company's "internal customers".

1.2.1.3. *Management of the supplier portfolio*

Calvi's [CAL 99] third main activity, management of the supplier portfolio, includes seven actions:

– measuring of the suppliers' performances;

– economic analysis of the suppliers;

– contractualization;

– auditing of the suppliers;

– negotiation;

– mapping of the suppliers' progress;

– order-processing and dispute-resolution.

Throughout the process of purchasing/supply, the buyer acts at the third, fourth and fifth stages, i.e. "choice of suppliers", "negotiation" and "order-issuing".

1.2.1.4. *Purchasing management*

Purchasing management is located at the strategic level of the purchasing process. It supports the activities of purchasing marketing and supplier portfolio management.

This process involves five elements:

– definition of purchasing policy (objectives, strategic guidelines, etc.);

– organization of the service;

– make or buy decisions;

– interactions with the other departments;

– measurement of the performance of the purchasing department, and budgeting.

1.2.1.5. *Purchasing maturity matrix*

The process of management of the purchasing function cannot be fully appreciated without reference to the overall strategy chosen. Its role is influenced by the level of maturity of this function within the company [VAN 96]. In this context, we speak of the "maturity matrix" principle, which enables us to model the purchasing function in a number of domains [POT 98]. From the recent works relating to this, we can cite the maturity matrix advanced by Bruel [BRU 04], Bruel and Petit [BRU 05] and Bruel again [BRU 07], which identifies five levels of maturity and highlights the increasing importance of the purchasing function in the organization's overall performance: "at stage 1, the Board of Directors practically only uses and pays attention to the productivity indicators and flow indicators. However, as the company grows and progresses vertically, the more complex the range of results indicators and processes should become. For instance, we would see the move from the notion of purchasing price to that of purchasing cost, and then to that of overall cost of acquisition. Stage 4 will justify indicators of upstream purchasing and joint performance indicators between purchasing and other functions, illustrating the breakthrough in nature relating to the passage between stages 3 and 4"

[BRU 05]. Each level is associated with seven internal and external actions (see Tables 1.4–1.8).

Level 1	"Basic" approach
General contribution	Supply management "Passive" purchasing (execution)
Purchasing policy/ internal levers	Focus on purchase order (PO) processing Technical categories approach to purchasing Pareto 20/80 with no segmentation
Supplier policy/levers	Approach to "supply" Negotiation for a certain amount > $XK
Processes/procedures /practices	Implementation and respect of a standard PO procedure No direct formal relations with the users
IDSSs (information and decision-support systems)	Basic bureaucracy Specific applications/ERP (Enterprise Resource Planning) modules for purchasing PO- and order-processing
Human resources for purchasing (collaborators)	Identification of buyers (best use of existing skills)
Internal and external communication (user relations)	Simple identification of internal customers

Table 1.4. *Purchasing maturity matrix: level 1 (source: [BRU 05; BRU 07])*

Level 2	"Internal" optimization of purchasing
General contribution	Upstream levers identified beginning to be used Reproducible purchasing processes
Purchasing policy/ internal levers	Formal segmentation of portfolio Definition of general purchasing policy Globalization Requirement planning

Table 1.5. *Purchasing maturity matrix: level 2 (source: [BRU 05; BRU 07])*

Supplier policy/levers	Competition and medium-term framework agreements (cat. A suppliers) (price, horizon, delivery conditions)
Processes/procedures /practices	Delegation (if OK) Formal rules for drawing up of specifications Follow-up with suppliers
IDSSs (information and decision-support systems)	Centralized internal purchasing database (capitalization on experiences from category-based purchasing) Supplier database
Human resources for purchasing (collaborators)	Definition and sharing of principles of HR management (job profiles, skills analysis) Buyer training plan
Internal and external communication (user relations)	Systematic needs analysis Committed contractual approach to "customers"

Table 1.5.(continued) *Purchasing maturity matrix: level 2 (source: [BRU 05; BRU 07])*

Level 3	**"Internal" optimization of purchasing**
General contribution	Widespread use of the upstream purchasing approach Recognized purchasing function
Purchasing policy/ internal levers	Segmentation (including out of production) Differentiated purchasing policy Analysis of purchasing markets External globalization Partial outsourcing Progressive internationalization
Supplier policy/levers	Systematic competition Construction of a (core) panel of suppliers Framework agreements/operational partnership (overall cost approach) Planned decrease in number of suppliers

Table 1.6. *Purchasing maturity matrix: level 3 (source: [BRU 05; BRU 07])*

Processes/procedures /practices	Purchasing manual Procedure sheets distributed and applied Internal audits Audits and suppliers' management (supplier quality assurance)
IDSSs (information and decision-support systems)	Generalized rational database (purchasing intranet) Dedicated market/sourcing database Use of Internet sites
Human resources for purchasing (collaborators)	Two-thirds of actors correspond to profiles Recruitment training Purchasing marketers put in place
Internal and external communication (user relations)	The procedures prescribed by the Procurement Manual distributed Purchasing policy distributed (CEO, users, partner suppliers) "Test sites" for specifications defined in communication with users

Table 1.6. (continued) *Purchasing maturity matrix: level 3 (source: [BRU 05; BRU 07])*

Level 4	**Optimization of the "buyer" company (all departments)**
General contribution	Integration of purchases into product design/business plan Contractual approach Participation of Purchasing Dept. in the "business" aspect
Purchasing policy/ internal levers	Action with upstream levers Standardization Design to total cost object approach Make or Buy Participation of Purchasing Dept. in project teams Technological sourcing International approach: e-sourcing
Supplier policy/levers	As above plus formalized progress plans Partnership on product development Generalized "panel"-based approach

Table 1.7. *Purchasing maturity matrix: level 4 (source: [BRU 05; BRU 07])*

Processes/procedures /practices	Periodic systemic audits – both internal and external Detailed procedures for product design and development: business (all Depts.)
IDSSs (information and decision-support systems)	Contribution to the database by all the actors not involved in purchasing: e-procurement
Human resources for purchasing (collaborators)	80% of purchasing actors conform to standard Training/information to users and purchasing advisors Upstream buyers put in place Internal mobility
Internal and external communication (user relations)	Test sites (upstream buying (next stage)) Active participation of Purchasing Dept. in project/business groups Formal communication plan "Motivating" actions

Table 1.7.(continued) *Purchasing maturity matrix: level 4 (source: [BRU 05; BRU 07])*

Level 5	**Optimization of the "buyer" company (all departments)**
General contribution	Company's purchasing performance "shared" Purchasing Dept. "leads" Purchases recognized and in demand
Purchasing policy/ internal levers	Full comprehension and appropriate use of all the levers Internal customers and advisors involved Purchasing strategy part of general strategy
Supplier policy/levers	80% of suppliers in partnership Co-development Innovation Strategic alliances
Processes/procedures /practices	Certification (e.g. ISO certification) of all purchasing processes (including purchasing and other services)

Table 1.8. *Purchasing maturity matrix: level 5 (source: [BRU 05; BRU 07])*

IDSSs (information and decision-support systems)	Generalized extranet and Internet system (use of Web applications)
Human resources for purchasing (collaborators)	Collaboration of all services Generalized contractual approach
Internal and external communication (user relations)	Generalized communication plan (all targets including suppliers on the panel)

Table 1.8. (continued) *Purchasing maturity matrix: level 5 (source: [BRU 05; BRU 07])*

Thus, the role of the purchaser is no longer limited – or, more accurately, can no longer be limited – to administrative routines. As highlighted in the French norm AFNOR NF X 50-128 (1990), he is an actor who is able to "instill in the suppliers a spirit of creative competition, to promote the suppliers' propositions and their suggestions to reduce costs and/or improve performances, and liaise with the other departments, informing them of any changes which affect their environment". His action extends over two levels [CAL 00], relating to the drafting of the contract which constitutes the pivotal point of the exchange:

– *ex-ante*, throughout the process of selection of suppliers, he verifies the knowhow, respect of commitments and reliability of the partner. In order to do so, he performs audits, establishes quotations, and considers the supplier's commitment by obtaining ISO standards. This painstaking work relates to the upstream-to-downstream supply chain;

– *ex-post*, by the monitoring of the relationship: this monitoring benefits a "win–win" relationship in which there is a transactional and relational dynamic based on trust.

Thus, the role of the purchaser expands, changing from simple participation to effective power "in each of the phases of the decision-making process" [CAL 00]. As Muller [MUL 04] points out, the purchaser has a significant positive effect on the performance of the purchasing department if four initial factors are in place. Thus, the purchaser has to reconcile efficacy (accomplishment of the task), effort (which the purchaser agrees to make in all the tasks relating to his function), skillfulness ("skills [...] to find the solutions that correspond to the needs formulated by the company") and precision of the role ("extent to which the perception of the individual's role, as formulated by the company's directors, is clear") – both quotes from p. 115 of [MUL 04].

1.2.2. *Internal or production strategic logistics*

The "need for speed" is accompanied by lean production[5] processes. This lean production is the preserve of the discourse of the logistics coordinator in production: maximization of the load of the equipment with running down of inventories of intermediary products [GIA 03; PAC 06]. Commonly, these two elements are rooted in the quality approach, with the example of Toyota. Thus, working with the basic idea of total elimination of waste, Ohno [OHN 89] combines individual talent and team tasks. Indeed, two main factors predominate:

– just-in-time production, which requires talent on the part of all those involved;[6]

– auto-activation of production. Auto-activation can be explained as the stopping of the production line in case of an anomaly,

5 Lean production logistics is practiced by maximum depletion or elimination of stock. It is similar to a just-in-time or *lean manufacturing* logic.

6 In the eyes of Bigras *et al.* [BIG 95], the just-in-time approach is explained by way of four fundamental principles which are:

– producing in exact quantities;

– producing uniformly;

– respecting individuals;

– retaining a long-term vision.

The objective being targeted is located on three levels: elimination of waste, constant improvement and elimination of unforeseen circumstances.

performed entirely at the responsibility of each of the factory operators.

Below, we present a number of decision-support tools/approaches which enable us to appreciate the various aspects of a production logistical strategy.

1.2.2.1. *A decision-support tool: the product/process matrix*

The product/process matrix is a decision-support tool for production. Developed in the 1970s, the product/process matrix characterizes the lifecycle of the production process [HAY 79a; HAY 79b; HIL 94; MIL 95; AHM 02].

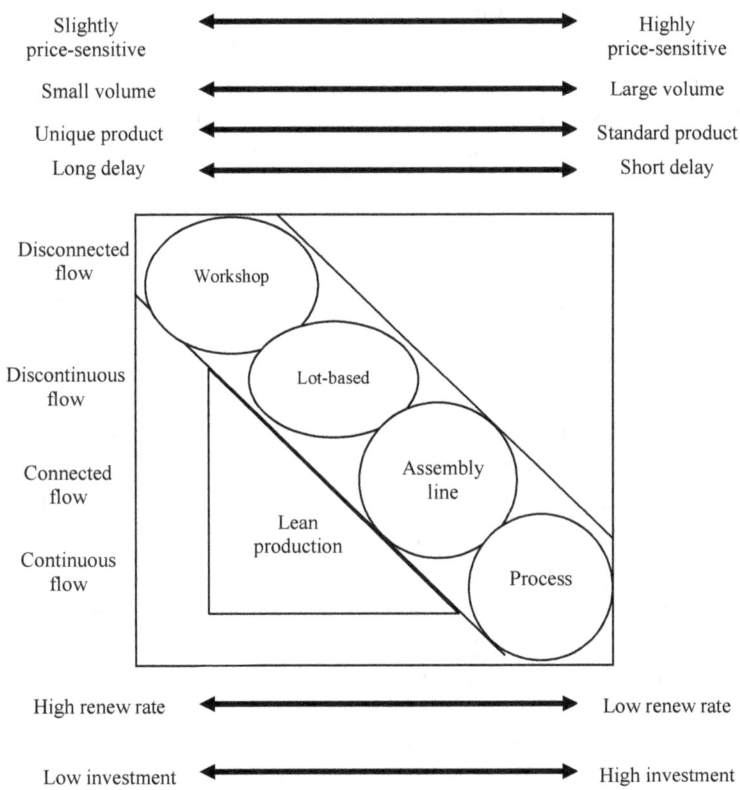

Figure 1.2. *The product/process matrix revisited (source: [BEA 08])*

According to Beaulieu and Landry [BEA 08], based on this matrix, the company chooses the best possible "alignment" with its partners, in order to reduce uncertainty and cater as best possible for the market being served. This market is arranged as shown in Figure 1.2.

In the top left-hand corner of the diagram, the factory functions with a fluid and flexible process. Conversely, in the bottom right-hand corner, it is faced with more rigid procedures. In general, an industrial sector begins its mode of production with a "disconnected flow", an emerging product with a small group of customers (a niche-based strategy [POR 82]). On the other hand, "continuous flow" results in a strategy of volume or domination by price, where the product becomes standardized and where the competitive edges are gained by a maximum of sales to a maximum of customers. However, above all, the process is chosen in relation to the market which is to be served. Drawing inspiration from this matrix, Ahmed and Schroeder [AHM 02] stress – firstly – a link with discontinuous- and connected-flow processes, and – secondly – the notion of *lean production* or *lean manufacturing*.

1.2.2.2. *One approach: lean manufacturing*

EXAMPLE.– "*Le lean, c'est de la compétitivité*" (Lean methods are tantamount to competitiveness) – *L'usine nouvelle*, 31 May 2012 (extract) [USI 12]:

> For its 2012 edition, the *Trophée des usines* (Factories Award) organized by *l'Usine Nouvelle* and *Insead* rewards two industrialists who have been able to launch ambitious projects of organization in order to enhance their competitiveness. And the winner is… lean manufacturing! More than ever this year, the method inspired by the Toyota Production System has become the number one weapon of industrialists for gaining in competitiveness. At Continental and Smart, the winners of our 18th *Trophée des usines*, it is indeed this method of industrial excellence which enables them to keep up production in France. In Foix (in Ariège), where Continental has its headquarters, or in Hambach (in

> Moselle), in "Smartville" (which covers Smart and its main suppliers), the very workshops breathe efficiency. Small islands where the flows are studied in order to reduce transition times, parts from suppliers which are put directly onto the assembly lines, workstations optimized with the help of expert ergonomists, production latency that adapts to demand… One can also sense the commitment of the operators. These two industrialists have put systems in place to reward ideas put forward by their employees. Continental's system was able to save the company nearly €550,000 in 2011.

Petitqueux [PET 06] establishes an overview of the implementation of the lean method in the industrial sector. Lean practice involves producing as accurately as possible. This approach was developed by Toyota and made public by the Massachusetts Institute of Technology (MIT) in the 1980s. The principles of the Toyota Production System (TPS) have become known by the umbrella term "lean manufacturing":

> Lean manufacturing is a generic name which denotes a production system originally developed by Toyota and used thereafter all over the world in all industrial sectors. Lean manufacturing is defined as a systematic approach aimed at identifying and eliminating all waste (non added-value activities) by constant improvement with a view to attaining industrial excellence. [PET 06], (p. 3)

Among the founding principles of lean manufacturing, we find [PET 06]:

– "industrial excellence, based on stable resources (products, costs, processes, trust, management, etc.);

– the notion of waste, with elimination of the 3Ms (Muri: irrational situations, Mura: variations, and Muda: non added-value tasks);

– *Kaizen*, by constant improvement;

– *Heijunka*, which advocates the division of the task of production into small lots, in order to limit inventories of "works-in-progress" and finished products;

– Just-in-time, which corresponds to the "ideal rhythm at which the production system should work in order to satisfy demand in a minimum period of time";

– *Jidoka*, the aim of which is to develop machines and processes which are able to detect any anomaly".

A number of tools are recommended in the context of lean manufacturing [PET 06]:

– Value Stream Mapping, which maps the processes, ranging from production to the whole of the supply chain;

> Value Stream Mapping (VSM) or Material and Information Flow Analysis. As it maps the processes and flows, VSM assesses the present state and determines a situation to aim for. This is one of the first operations to be performed when we are using an approach to transform these processes. VSM opens the door to the use of the following (more operational) tools.

– the operations diagram, which focuses in particular on non-added-value tasks, with a view to doing away with them;

– the 5S approach, which aims for general standardization by way of sorting (*Seiri*), setting in order (*Seiton*), systematic cleaning (*Seiso*), standardizing (*Seiketsu*) and sustaining (*Shitsuke*);

– delayed differentiation, which involves adjusting push-flow (standardization of the product) and pull-flow (personalization of the product) in order to gain time on production and therefore delivery;

– the SMED (Single Minute Exchange of Die) method, which requires the production line to be changed from one product to the next in less than a minute;

– the assembly line, where a thought relates to decreasing wastage such as waiting times, transport or indeed movements;

– ergonomics of the workstation;

– synchronization of the stations with the Takt Time, i.e. the quest to limit bottleneck situations;

– unitary production flow, where each upstream operation continuously feeds into the next operation;

– *Poka-Yoke*, which consists of determining any technique which limits errors owing to the weariness caused by standardization;

– *Kanban* or label, with a view to lean manufacture;

– "little train", which refers to the time gained by the movement of the warehousemen, using its warehouse trolley;

– "short-delay encouragement", which puts in place frequent reports, the goal of which is to stay abreast of the desired and actual levels of performance.

1.2.3. *Upstream or distribution strategic logistics*

For many companies, their major concern is to guarantee durability whilst ensuring that the customer, who has become "volatile", proves to be a partner in the long term. The satisfaction of the end consumer is one of the major concerns for any company, and therefore of downstream or distribution logistics.

This is one of the reasons for which this field of thinking was one of the first to be the subject of research [COL 88]. In its practices, distribution logistics exhibits parallels with marketing. On this link between marketing and logistics, Senkel [SEN 02] details four main evolutions, highlighting a thought process which has moved from the opposing interests of marketing and logistics (marketing is interested in the customer, whilst logistics coordinates activities so as to minimize the cost-to-market of the product) [CHR 85], to a convergence between the two fields (marketing and logistics both contribute, in parallel, to customer satisfaction) [RIN 89]. Thus, we see the interface between marketing, logistics and production [LAM 94] and, finally, come to consider logistics as a necessary condition for the elaboration of a marketing strategy [AUR 97].

1.2.3.1. *The Efficient Consumer Response approach*

In downstream logistics, it is the company's objective to satisfy the customer, rather than *vice versa,* so here we do not find evaluation matrices such as those demonstrated for upstream logistics. In fact, downstream logistics is a key part of the very definition of logistics in the general sense: delivering the right product, at the right time, in the right amount and at the right place. For this reason, it is primarily the approach known as ECR (Efficient Consumer Response) which is favored in this link in the supply chain.

ECR is testament to an increased willingness for logistical cooperation [BON 99; BON 05]. Founded in the United States in the 1990s, ECR "echoes the idea whereby it is in the interests of the members of a channel to cooperate in order to improve its efficacy, evaluated on the basis of criteria such as the improvement of the service rate or the lowering of the number of breakdowns" [PAC 97] (p. 67). While the implementation of this general model of partnership saves costs all along the chain, which runs from the food distributor to the industrial supplier [VAN 98], Des Garets [DES 00] points out that the primary goal of ECR is the application of the following five basic principles:

– "(...) to provide the consumers with an added value in terms of products, quality, range, availability and rate of service, whilst reducing the cost of operating the supply chain;

– to adopt a logic of cooperation rather than one of transactional confrontation;

– to facilitate [the marketing, logistical and production choices] by way of rapid and reliable information exchange through the EDI (Electronic Data Interchange);

– to guarantee the availability of the right product at the right time;

– to distribute [by a measurement of overall efficacy] the benefits obtained by the partners" (p. 113).

Hence, ultimately, we find significant implicit convergences between the expectations of distribution logistics and overall logistics. Coordination by one and by the other underlies a process of interactions between those involved in the operations chain from downstream (the demand to be served) to upstream (supply), i.e. the manufacturing company, its suppliers and its logistics providers, but also its distributors who, by creating common distribution platforms and combined road networks, play an important role in the making of large-scale savings.

1.2.3.2. *From e-commerce to u-commerce*

E-commerce, for electronic commerce, is part of the business to consumer relationship. In his own right, the consumer defines and places his order via his computer. Thus, FEVAD (*Fédération des Entreprises de Vente A Distance* – Federation of Remote Sales Companies) reports a 22% increase in online sales in France between 2010 and 2012. In other European countries, we note the following increases: Germany: +17%; Belgium: +24%; Denmark: +14%; Italy: +19%; and the Netherlands: +9%. To cite another statistic reflecting this trend, in 2011, 43% of private customers in Europe said they had made purchases over the Internet. In terms of logistics, this has consequences for the ways in which orders are prepared and delivered. As regards the delivery method, the logic comes under the same issue as urban logistics, which is discussed later on in this book. With regard to the means of preparation of orders placed online, two basic models are chosen [PAC 08; DUR 10]:

– preparation on a purpose-specific site located at a warehouse. This type of preparation is used when the trader's activity mainly rotates around digital materials;

– preparation in store, where electronic activity plays a lesser role.

Thus, increasingly, logistical approaches are being devised to cater for the RoPo (Research Online, Purchase Offline) effect [VIN 12]. Amongst these approaches, we can cite the drive points at Carrefour stores in France, which offer delivery "to the trunk of your car" for heavy purchases, so as to leave you free for "leisure shopping" in the store. In summary, a great deal of in-depth thinking has gone into making e-commerce as close as possible to the new telecommunications technologies, and making the consumer co-authors in their own purchases. Here, we speak of u-commerce, meaning ubiquitous-commerce, where computer technology becomes omnipresent in the consumer's act of buying.

EXAMPLE.– "*L'e-commerce, c'est dépassé*" (e-commerce is a thing of the past) – C. Vincent, *Enjeux Les Echos*, June 2012, p. 32-41 [ENJ 12].

> M, T, F, U… the many avatars of e-commerce. Yesterday, it was all so simple. We had e-commerce. Yet now, good old "electronic commerce" is no longer sufficient to express the multiplicity of the digitalized mercantile world. Recently, we have begun to witness the triumphant entrance of m-commerce – "mobile" commerce, conducted through a smartphone or tablet. Anyway, some would draw parallels between these devices – often used as a second screen in the home – and the Internet-connected televisions which are beginning to break through, and form the family of "t-commerce", which combines TV and tablets. An untimely *t* which enters the arena to compete with another so-called t-commerce: that conducted *via*… Twitter. For let us not forget social networks, and Facebook in particular. Thus, we have f-commerce, so we can buy things that our friends have "Liked". Finally, let us pay homage to King Google. G-commerce would be guaranteed a bright future by its exegetes. Let us simplify the situation a little. Ladies and Gentlemen, we give you… u-commerce: ubiquitous commerce. Or everything is in everything, everywhere, all the time.

1.2.4. *Strategic transport management*

The last three sections have dealt with so-called intra-organizational logistics (although this still entails links with the outside world). However, this type of logistics, which is present in every (industrial) firm, requires us to use forms of transport (road, sea, river, air, rail) in order to ensure the flow of goods and documentation.

In this section on the notion of coordination of modes of transport, we focus particularly on transport pooling.

1.2.4.1. *The extended transport service*

Long before the notion of transport pooling, Colin and Paché [COL 88] pointed out that the question of an extended transport service was arising. Such extended transport requires skills which go beyond the "simple" transport of goods.

The extended transport service [COL 88] is divided into seven components:

– transport operations (supply; inter-factory transport; physical distribution to depots, platforms, points of sale) which include consolidation/deconsolidation operations, organization of delivery rounds and rent of vehicles, with or without a truck driver, for a long or short period;

– auxiliary transport operations, with transit, customs duty and transport commission;

– physical distribution operations, which include 17 actions – namely:

- various handling operations and sometimes management of a pallet pool,
- receipt of the goods,
- quality and quantity control,
- placing on reserve,
- withdrawal,
- division,
- assortment and allotting[7],
- labeling and price marking,
- creation of barcode labels for the management of internal flows and/or transport,

7 Allotting consists of grouping packages together by their destination.

- reconditioning,

- preparation of order,

- weighing,

- packaging, outer packing (palletizing, film wrapping and shrink-wrapping[8]),

- preparation of cargo,

- pre-billing,

- dispatch,

- final delivery;

– management operations with:

- warehousing, stock management,

- computerized management of products by the carrier, processing of orders to be delivered,

- management of expiration dates,

- sometimes, and increasingly, management of stocks of raw materials, supplies, semi-finished products and finished products, for industrial units working on the "lean manufacturing" model and which do not have any storage capacities; and connection of the supplier's computer system with that of the dispatchers and addressees;

– commercial operations, which are based on six actions:

- billing,

- merchandising, line management at the points of sale,

- sometimes installation and after-sales service (maintenance) for certain products,

- offices and logistical and computer resources being made available for the carrier at the transporter's headquarters,

- construction of modular and hierarchized rates for the different modular components of the services offered,

- management and regulation of complaints;

---

8 Shrink-wrapping consists of wrapping groups of (usually identical) products in a cocoon of retractable film.

– "production finishing" operations, which include the final construction of testing banks (for certain mass-consumption manufactured goods such as bulky home electrical goods);

– advice and logistical engineering operations, in transport and physical distribution, sometimes in stock management, and less frequently even in production management, defining means and procedures to be implemented to optimize logistical operations (for the dispatcher, the supplier, the addressee).

It can be said that since the creation of this list in 1988, these components have changed little. Certainly, the creation of barcode labels (third point) is still common practice, but it has evolved toward the use of RFID (Radio Frequency Identification) tags, depending on the value of the products. Similarly, we are witnessing an enrichment of certain tasks, which tend towards advice-giving and accompaniment, as happens in the case of logistics services.

1.2.4.2. *Toward a quest for transport pooling*

Nowadays, in the writings about collaboration in transport management, we often find mention of transport pooling. Transport pooling is defined as:

> "a type of inter-company collaboration characterized by a pooling of resources between two or more actors (who may or may not be competitors) with the aim of better rationalization of transport flows. By rationalization of transport flows, we mean any means (economic, technological, human, organizational, etc.) by which one can improve the delivery of a good or service to the point of its final consumption" [GON 10A; GON 10B].

For these authors, there are two possible modes for the operation of transport pooling:

– multilateral collaboration, which consists of pooling the resources of a multiple actors, with or without an intermediary party;

– partnership of transporters whereby, through a cooperative association or society, a computer platform is put in place to federate supply and demand.

1.2.4.3. *Urban logistics spaces*

Finally, in transport management, we must point out the importance of urban logistics, or "last-mile logistics". Thus, we shall discuss the pilot project set up in France by the Sustainable Development Ministry, known as *espaces logistiques urbains* ((ELU) urban logistics spaces). These ELUs constitute logistical installations whose aim is to improve the transport of goods between the highway network and the destination site, and between cities and their (more or less distant) peripheries [BOU 06]. There are different kinds of ELU:

– on the outskirts of urban areas, we have *zones logistiques urbaines* ((ZLU) urban logistics zones) which bring a number of transportation companies. The aim of the ZLUs is not to mutualize the flows of goods between these companies, but rather they adhere to the regional logistical policy. Consequently, the presence of a ZLU does not alter the delivery and collection practices of any of the companies involved;

– usually a few miles from the city centers, we have *centres de distribution urbain* ((CDU) urban distribution centers), which are platforms for consolidation/deconsolidation operations, the intention behind which is to manage flows bound for densely trafficked areas;

– in the neighborhoods and the city streets, we have *points d'accueil de véhicules* ((PAV) vehicle stopping points or loading bays). These are infrastructures which take the form of mini logistical platforms, where part of the road is set aside for goods transport vehicles to park. In this category, we observe a number of sub-categories such as:

- *espaces logistiques de proximité* ((ELP) local logistics spaces),

- *points d'accueil de marchandises* ((PAM) goods-receiving points),

- delivery areas.

Of these PAVs, the ELPs might be able to alter logistical policies in order to enhance the delivery of the actors involved. Thus, they may offer services similar to those of a CDU, but on a smaller scale;

– *boîtes logistiques urbaines* ((BLU) urban logistics boxes), which are installed at the level of a street or even a building. These are small structures, either moveable or fixed, which offer an interface and a temporary storage space for goods so as to optimize the performance of delivery rounds. This category includes purpose-built hatches, lockers and new points for collection of delivered goods close to the place of consumption.

Name	Creation date	Number of people	Turnover
Adrexo Colis	1979	23,000	€330M
Chronopost	1985	3,500	€665M
Coliposte	1996	7,500	€1.4Bn
Colizen	2008	20	€1M
Exapaq	1995	2,000	Undisclosed
Fedex Express France	1985	2,700	€38Bn
France Express	1971	4,200	€306M
Kiala	2001	110	€47M
La Petite Reine	2001	70	€1.5M
Mondial Relay	1997	550	€100M
Pickup Services	2000	70	€9-10M estimated
Relais Colis	1983	750	Undisclosed
Morin logic	1954	380	€34M
SLS – GCI	2010*	750	€82M
TNT Express France	1973	5,000	Undisclosed
Top Chrono	1984	140	€10M
UPS Europe	1976	40,000	€40.9Bn

* date of merger between SLS and GCI

Table 1.9. *The actors in last-mile logistics in France (source: data taken from Logistiques Magazine, May 2011, no. 259 [LOG 11])*

As we can see, the different ELUs do not serve the same function or have the same scope for action. For our discussion, we shall focus in particular on the consolidation/deconsolidation platforms, the CDUs, which have an appreciable effect on the organization of the transport of goods in last mile logistics.

1.2.4.3.1. *The important role of CDUs*

Rationalization of the flows of goods is founded in the notion of an urban distribution center (CDU). The CDU is generally appreciated as a platform where (high-tonnage) transport flows coming into a city are concentrated, to be put into lighter vehicles, which emit less pollution and have less difficulty in accessing denser areas.

There are three types of CDU [BRO 05]:

– "private" or "semi-private" CDUs. Their purpose is essentially economic, and they form part of the development strategy of the operator or carrier;

– "mutualized" CDUs, which are generally associated with a service promoted and/or supported by the authorities. The aim behind the mutualization is to pool resources in order to manage flows bound for the city as closely as possible;

– "specific" CDUs are associated with particular activities. They may be temporary (platforms attached to work sites or large-scale moves) or permanent (airports, seaports or river ports). Yet, unlike the previous two types of CDU, their activity may not necessarily be geared towards urban distribution.

In itself, this last-mile logistics appears to be an important factor in companies' thinking. Indeed, it represents the trickiest and most costly link in the supply chain: trucks above a certain tonnage cannot enter urban areas, there are frequent stops involved, which drives up the costs, and there is the issue of whether the buyer will be there to receive the goods. In one of the rare examples of success of this strategy, Morana and Gonzalez-Feliu [MOR 11] (p. 24) stress two criteria which all private companies need to consider:

– "co-funding [public-and-private] of vehicles and use of one's own infrastructures, in order to minimize the costs of implementation of the service, using regional and European funds for the development of 'green' strategies;

– operational management, which combines outsourcing of common operations with a highly dynamic commercial strategy, so as to seek out new customers at any time. In addition, the use of an information system accepted by the main transport operators [...] means that a diversification of services can also be envisaged".

EXAMPLE.– "Denis Badré: *Réapprovisionnement du centre-ville, un regain d'énergie ?*" (Restocking City Centers: energy recovered?), *News Press*, 19 October 2010 [NEW 10].

The city of Bordeaux has been awarded the first "Urban Logistics Prize" for its CDU. Last-mile delivery is done by electrical vehicles or delivery tricycles.

EXAMPLE.– "*Création du Centre multimodal de distribution urbaine (CMDU) à Lille*" (Creation of the multimodal urban distribution center in Lille), *Transports & Logistique*, April 2012, p. 26 [TRA 12].

On 27 March 2012, the GIE-CMDU was formed when Ports de Lille, CCI Grand Lille, Generix Group, MIN de Lomme, TLF and Véolia Propreté joined forces to test out the urban logistics of the future, in Lille.

The aim is to reduce the problems thrown up by the transport of goods in a city environment, and primarily road congestion and pollution. Using the infrastructures of the city of Lille, located near to the city center, the CMDU will receive goods by sea, road or rail. After optimization of the delivery rounds, the goods will be delivered by the transporters themselves to their recipients, with preference given to clean modes of transport – particularly electrical vehicles.

1.2.5. *The increasingly strategic role of logistics providers*

In talking about the extended transport service, it is natural to speak of the role of logistics services. This is the subject of many discussions: first, in terms of the service provided; then in terms of the place of that service within a "conventional" supply chain approach (definition, role, power, etc.). Nevertheless, logistics services are garnering support. In evidence of this, we wish to hold up the appreciation of this activity on several levels. Thus, Artous [ART 03] presents the evolution of the role of a logistics provider. At that time, four levels of expertise were referenced:

– 1PL: for First Party Logistics (Provider), which consists of sub-contracting of transport or warehousing;

– 2PL: for Second Party Logistics (Provider), which includes both the outsourcing of transport and warehousing;

– 3PL: for Third Party Logistics (Provider), logistics providers who handle the various different flows stemming from logistics and who have at their disposal means of transport, warehousing, etc. to carry out this function;

– 4PL: for Fourth Party Logistics (Provider), who are providers whose service of handling of logistical flows takes place only at an informational level. The 4PL plays the role of a link between the company, the market and the other providers.

Less frequently, in the world of logistics providers, we also speak of 5PL, for Fifth Party Logistics (Provider), whose role is to coordinate the activities of the various subcontractors, but also to design new computer-based logistical solutions[9].

Finally, LLP – Lead Logistics Provider – who are on the boundary between 3PLs and 4PLs, as they provide a service for goods that they themselves own, but also for those of other companies. Thus, as Fulconis *et al.* [FUL 11] put it, "the time has now come to view

9 www.olf.gouv.qc.ca/ressources/bibliotheque/dictionnaires/terminologie_logistique/prestataire_services_logistiques.html.

logistics provision through a very different lens, enabling us finally to see the exceptional potential for technical and organizational innovation that logistics providers offer" (p. 257).

N°	Name	Logistical turnover	Warehouse surface area	Ware-houses	Logistic al staff
1	Kuehne+Nagel	€527M	1,500,000m²	58	6,000
2	Norbert Dentressangle	€517M	5,300,000m²	193	13,836
3	Géodis	€450M*	Undisclosed	Undiscl.	Undiscl.
4	DHL Supply Chain	€400M	1,000,000m²	85	4,500
5	STEF-TFE	€361M	4,052,000m²	85	3,057
6	Bolloré Logistics	€300M	506,000m²	84	60
7	FM Logistics	€291M	>2,000,000m²	>70	12,000
8	ID Logistics	€191M	1,975,700m²	69	6,250
9	Soflog-Telis	€155M	300,000m²	50	1,200
10	Mory **	€136M	574,000m²	49	67

* estimation; ** In 2011, Mory's logistical infrastructure was bought up by ID Logistics.

Table 1.10. *Logistics providers – taken from the breakdown of the Top 10 in France (source: Logistiques Magazine, no. 255, 2010 [LOG 10]*

1.2.6. *The strategic approach of traceability*

In order to firm up the intra- and inter-organizational connections upon which SCM is founded, we cannot overlook the role and weight of traceability in the pursuit of logistical exchanges. The notion of traceability arose in the wake of the food crises and has been growing in importance ever since in all business management. The ISO 8402 standard from 1994 defines traceability as follows:

> Traceability is the aptitude to find the history, the use or the localization of a product by means of recorded identification.

The goal behind traceability is to be able to withdraw the product in case of a problem – as quickly as possible, and only for the contaminated lot, so that the problem does not affect all articles of a product. Traceability also has a part to play in the war on counterfeiting, because *tracking* (real-time monitoring of the flows) and *tracing* (memorization of the product) make it possible to verify its origin.

The functions of "*tracking*" and "*tracing*" (taken from [ROM 00]):

> The function of "tracking" consists of being aware of an object's location at any given time *t*. The objective associated with this [...] function of traceability is twofold:
>
> – on the one hand, the aim is to demonstrate reactivity in the face of a problem which transpires, such as the stopping of a defective product, and to be able to rapidly determine the cause of the problem and the exact location of the lot in question;
>
> – and, on the other hand, the aim is to have better control over the whole of the supply chain, by way of real-time monitoring of the physical flows. This control involves, e.g., better stock management, better quality in order preparation, etc.
>
> In order to implement the function of 'tracking', we need to:
>
> – identify the objects by way of a coding system and gather data by means of an appropriate information system: most often, this is done automatically, by a barcode system with an optical reader or electronic tags with a radio-frequency system;
>
> – have a tool which is able to observe and analyze the flows in real time and communication information about them. The means for this may range from the simplest to the most advanced tools: telephone, fax, EDI, a software package, the Internet or indeed radios, or onboard computers installed in the vehicle transporting the object being tracked, or a satellite location system.
>
> The function of "tracing", for its part, involves providing an overall view of all the flows in which we are

> interested, i.e. retracing the prior history of the object's actual circulation. The objective here is to be able to go back to the origins of the products, and monitor each stage of their production. This requires us first to have recorded the data produced by the 'tracking' function, and then to have a system capable of combining these data so as to reconstruct the overall image of the flow in question.

Traceability is considered to be a strategic management approach, requiring three main stages: the organization of the project, the implementation of the approach and the use of the information [KAR 10]:

– the organization of the project is divided into seven "standard" stages, which:

- define the context of the project,
- puts the working structure in place,
- determines the requirements,
- evaluates what is already in existence,
- suggests solutions,
- executes the project effectively,
- ensures tracking and evaluation;

– the implementation of the approach requires four stages, i.e.:

- identification of the products,
- management of the connections,
- recording of the data,
- communication.

The link between the organization of the project and the implementation of the approach is made by ICT with barcodes, RFID, EDI, etc.:

– the use of the information includes:

- monitoring of the product's safety, origin and integrity,

- strengthening of the company's competitive advantage,

- constant control and evaluation,

- redefinition of the relationships between the different actors whom traceability concerns.

For the same author, there are four dimensions which define the strategic approach to traceability:

– the characteristics of traceability;

– the characteristics of the environment;

– the characteristics of the organization;

– the transversal factors.

For each of these aspects, it is possible to specify the factors which will help the adoption of a traceability approach.

1.2.6.1. *The characteristics of traceability*

In order for traceability to succeed within the company and with its partners, the approach needs to exhibit a certain number of characteristics. These characteristics of traceability are based on three factors:

– observability, which is the extent to which the results of the traceability are visible and observable;

– compatibility, which involves common values, requirements and past innovations. The past innovations here refer to partnerships between actors and to the practice of a quality approach;

– the relative advantage combines the following ten elements:

- the survival of the company, in the sense that no company can survive unless it adopts a traceability approach,

- the relations between actors, founded upon transparency and trust, selection of suppliers, coordination and cooperation,

- the internal organization, which involves management of information and of documentation,

- the image of the country,

- the image of the company,

- the time gain,

- the development of the company,

- control and evaluation, which involve having an overall vision, management of risks through customer complaints, targeted withdrawal and causal links between complaints and withdrawal,

- competitive advantages, which involve customer satisfaction, cost reduction and consumer confidence,

- quality control.

1.2.6.2. *The characteristics of the environment in traceability*

The second dimension in the traceability approach takes account of three factors:

– the role of the stakeholders, in the broadest sense of the term. Here, particular attention will be paid not only to the staff (adoption and understanding of the approach) and the upper management (project owners), but also to external actors such as the supporting organizations which facilitate the implementation of a traceability approach: the State, the consumers (national and international), the media, the suppliers, etc.;

– the champion, who is the person that uses his/her position and sway within the organization to influence an organization's ability to participate in new technological and/or organizational innovations such as traceability;

– the competitive environment which corresponds – all other things being equal – to the high level of competition on the market.

1.2.6.3. *The transversal factors of traceability*

Finally, the third dimension examines three factors:

– the motivation to put in place financial subsidies (internal and external, in the case of support from outside organizations) for the success of this approach;

– the level of partnership, which corresponds to the degree of maturity of the partners in traceability within the same supply chain;

– the relations between partners, whereby cooperation is considered to be of prime importance.

Often, traceability is seen as a tool, a means for supply chain management [ROM 00; FAB 01; COL 05]. Sometimes, it is viewed as an approach in its own right [VIR 04; PEL 07]. The results presented here are the feedback on doctoral research into the understanding of traceability in a country just beginning to practice this approach. It should be noted that all the elements cited tend to stress the fact that traceability should be perceived as an approach in the same sense as supply chain management.

1.3. Information, information systems and ICT: an aid to the success of intra- and inter-organizational connections

Intra- and inter-organizational connections such as those seen above require increasing amounts of information in order to function. For this reason, it is necessary to put information systems (ISs) in place. An information system is defined as "an organized set of resources: hardware, software, HR, data and procedures, to acquire, process, store and communicate information (in the form of data, texts, images, sounds, etc.) in organizations" [REI 95] (p. 67). The goal is to achieve governance of the ISs, with the aim of "encouraging actors to rethink and alter their practices of coordination and learning, to actively contribute to an innovative dynamics which requires trust, information sharing but also confidentiality" [ELA 08].

In logistics, there are many software tools which serve many different functions. Here we shall present some of the most common of these tools.

For Fabbe-Costes [FAB 00], the typology of SCM technologies can be thought of by way of three axes:

– decision-support technologies, which organize decisional flows into a coherent form. These include APS (Advanced Planning and Scheduling system), SCM tools, databases, datawarehouses, etc.;

APS, standing for Advanced Planning and Scheduling system, is a software package relating to production management. It draws the balance between the customers' demands and the companies' capacity to cater for these demands.

– interface technologies, which set up monitoring of cooperative commitments and communications between each partner in the chain. These technologies include EDI (Electronic Data Interchange), the Internet, intra- and extranet, ECR (Efficient Consumer Response), GPA (*gestion partagée des approvisionnements*, or shared supply management), call centers, etc.;

EDI is a tool which transmits structured data between computers, in accordance with pre-established and standardized messages (see the appendix to this chapter for a view of the elements in the construction of an EDI guide).

GPA: "*Gestion partagée des approvisionnements* (GPA) is a form of collaboration between distributors and their industrial suppliers, which is rapidly gaining popularity at present. GPA is based on intense electronic data interchange (EDI). The distributor, or the logistical service provider entrusted with the management of the warehouses, sends data on a daily basis to the supplier, concerning outgoing stock for the categories of product in question. These data enable the supplier to make informed predictions and create a suggested order which is sent to the distributor's purchasing center for validation. Hence, the flows are "pulled" by the demand. This technique is highly appropriate for mass-consumption products which are frequently delivered (one or more times a week)." [FEN 05] (p. 206).

In the same vein as GPA, we also have collaborative supply management, better known by the acronym CPFR, for Collaborative Planning, Forecasting and Replenishment. In comparison to GPA, CPFR includes promotional campaigns.

Similarly, at this level we find the concept of Vendor Managed Inventory (VMI) or Supplier Managed Inventory, whereby the supplier manages the stock on behalf of their client.

Finally, by extension, we speak of GMA (*gestion mutualisée des approvisionnements* – Mutualized Supply Management) which is defined as "a means of supply management where a number of industrialists agree to make combined deliveries, departing from the same logistics site (warehouse or platform), one or more distributors, notably so as to optimize storage costs (reducing the number of storage sites in the industrialist-to-distributor supply chain) and transport costs (extent to which the vehicles are filled and the number of rounds" [LIV 06].

– piloting technologies which monitor, evaluate and integrate the major processes in Supply Chain Management. These include ERP (enterprise resource planning), and Manufacturing and Logistics Execution systems.

ERP is a program which handles the operational functions of an organization, such as finance, accounting, marketing, production, etc. Essentially, its objective is to deal with the concerns of analytic division by function and by monitoring of the flows.

In a continuing train of thought, Fabbe-Costes [FAB 07] lists the systems which are able to take account of transport as a link in logistical chains. These systems are distributed around three axes:

– document exchange systems, which include the transportation order, the delivery note, the transport receipt, the bill of lading (B/L)[10], the packing list, etc. The tools used here are automatic fax, EDI, the Internet (email, Web-EDI, online communication);

– communication systems including onboard radio, embeddable/portable terminals, telephones, mobile telephones, the Internet (email, online communication) and multi-functional portable terminals;

– traceability systems, including identification/coding, marking, electronic readers, tags, voice systems, recorders and memorization.

> Identification/coding is established by two means: a barcode, which has a unique use, and RFID (Radio Frequency Identification) which can be used many times.
>
> The barcode system dates from 1952. It was created by two American engineers, N.J. Woodland and B. Silver. To this day, it remains the most widely used system for product identification and coding.
>
> RFID dates from the late 1940s. It was initially used by the Royal Air Force in the Identify Friend or Foe (IFF) system. Its use in industry dates back to the 1970s. Its flexibility (multiple use of each chip) gives it an advantage over the "single use" barcode system. However, its cost as yet remains an obstacle to its expansion.

Besides the examples given above, other software tools help improve the efficiency of a supply chain. One might cite:

– CRM – Customer Relationship Management;

> CRM can be understood as an approach as well as a software tool. Its keyword is *optimization*: optimization of customer relations, whether present, future or prospective.

10 The bill of lading (B/L) is the document which embodies the maritime transport contract agreed between the cargo handler and the maritime transporter. It is also a certification of the ownership of the goods.

– DRP – Distribution Replenishment Planning – which helps manage inventory of stocks of finished products, whilst taking account of the complexity of distribution networks;

– MES – Manufacturing Execution System. This term denotes systems which provide information in real time about the execution of manufacturing orders. The objective is controlling of the production process, from its beginning to the delivery of the finished products. Thus, the MES seeks to optimize production activities;

– MRP – Material Requirements Planning – was created in 1975 by Orlicky, and is a program for managing production and supply which calculates the net requirements with infinite capacity. It corresponds to a set of techniques using the nomenclatures, the stock status and the production control program, so as to calculate the material requirements. It is viewed as a planning technique intended to establish and maintain correct due dates. Its extension is MRP2, standing for Manufacturing Resources Planning;

– SCE – Supply Chain Execution – denotes a category of software packages for the purpose of operational management of the supply chain. It includes:

- WMS – Warehouse Management System – which is an information system for preparation, tracking and execution of activities in warehouses,

- TMS – Transport Management System;

– SRM: Supplier Relationship Management, which involves everything relating to the handling of the supplier relationships.

1.4. Conclusion

We are able to gain an appreciation of the economic aspect of sustainable supply chain management from the publications associated with supply chain management. In particular, it is the intra- and inter-organizational connections which are highlighted. Thus, companies are no longer isolated, and can go further in information-sharing and exchange. They are supported in this aim by information and communication technology, which represents crucial support mechanisms in any supply chain.

The goal of this chapter was to present a number of strategic tools (matrices, guides, approaches, etc.) which accompany the economic implementation of sustainable supply chain management. Here, the intention is to show how this strategy fits into an upstream approach to the implementation of sustainable supply chain management, by setting out the (few) focal elements necessary to ensure its success.

1.5. Appendix – technical specifications for electronic data interchange

Arbaoui and Morana [ARB 04] propose the construction of a set of technical specifications for EDI, specifying and/or adding elements of construction drawn from management sciences [ROL 99] and process engineering [ARB 03]. As such, the basic structure of these technical specifications includes eight clauses.

1st clause: the aim of the description of the problem is to introduce the context, give background for the service and specify the "malfunction" needing to be dealt with:

– taken from management sciences, two elements in particular need to be taken into account:

- the organizational/resources potential,

- the organizational/activities potential;

– taken from process engineering, two elements in particular need to be taken into account:

- the description of the process in a formal/semi-formal notation, and construction of reusable models,

- the environments for the use of the models when commercialized (Workflows, Merise).

2nd clause: the constraints section aims to present:

– material constraints, including:

- taken from management sciences, two elements in particular need to be taken into account: (1) the technical/resources potential and (2) the technical/activities potential,

- taken from process engineering, the following elements in particular need to be taken into account: graphic languages to model the processes in terms of tasks, roles and objects;

– social constraints (adaptation at the level of the users), including:

- taken from management sciences, two elements in particular need to be taken into account: (1) the human/resources potential and (2) the human/activities potential;

– access security constraints, including:

- taken from management sciences, four elements in particular need to be taken into account: (1) the nearby environment, (2) the measurement of the pressure on the company, (3) the move from an organizational form to a market logic and (4) the evolution from partner relationships to a concept of electronic hierarchy,

- taken from process engineering, the following elements in particular need to be taken into account: the modeling languages and implementation environments, taking account of the geographic distribution of the processes and the interoperability of the systems.

3rd clause: the functionalities clause aims to describe the existing and desired functions:

– taken from management sciences, four elements in particular need to be taken into account: (1) the logistics, (2) the organization, (3) the relationships with the surrounding environment and (4) the company's strategy;

– taken from process engineering, two elements in particular need to be taken into account: (1) evaluation and improvement of the processes and (2) analysis, control and quality of the processes.

4th clause: execution time.

5th clause: the "who does what" at the level of the project manager and the contractor:

– taken from process engineering, one element in particular needs to be taken into account: knowledge of the processes and modeling capabilities.

6th clause: confidentiality:

– taken from process engineering, one element in particular needs to be taken into account: the mechanisms for the implementation of the distribution and cooperation. The objective is to manage accesses and configurations, and to synchronize the data.

7th clause: the modes of reception, which encompass the pre-recipe, the definitive recipe, the guarantees, unpacking/installation and security.

8th clause: the particular clauses with after-sales service and training.

Chapter 2

The Environmental Aspect of Sustainable Supply Chain Management

2.1. Introduction

Current practices demonstrate the important place that waste-management occupies in existing organizational policies. The regulatory framework as it currently stands is oriented toward environmental awareness, as demonstrated by the decree on "Waste Electrical and Electronic Equipment". Yet companies are seeking to go beyond "simple" waste-management, and are starting to think about waste-re-engineering, which begins at an early stage in the circuit of production of the product.

In the movement of reflection on green logistics, we present Srivastava's summary publication [SRI 07] on Green Supply Chain Management (GrSCM). Srivastava defines GrSCM, which first surfaced in the 1990s, "*as integrating environmental thinking into supply-chain management, including product design, material sourcing and selection, manufacturing processes, delivery of the final product to the consumers as well as end-of-life management of the product after its useful life*" [SRI 07, p. 55]. He consolidates this definition with three elements: the importance of GrSCM, green design and green operations.

2.1.1. *Integrated logistics support*

Before discussing these three points, it is interesting first to talk about integrated logistics support (ILS), which dates from the 1960s and has a part to play in "green" thinking.

> Logistics can [...] encapsulate the whole of the life-cycle of a manufactured piece of equipment, beginning with the design phase: thus, it is an "integrated logistics support" approach, which is essentially to be found in hi-tech industries (armaments, electronics, aeronautics, space). Its basic principle is relatively simple. In order to be able to manage the events likely to occur over the course of the life-cycle of a piece of equipment (particularly maintenance), we need to take these events as sources of feedback on the design itself, so that the machine does not pose problems, either during its manufacture or its operation (modular design facilitating the maintainability as well as the production of a machine, availability and supply of replacement parts within periods calculated even at the design phase...) [COL 88] (p. 49).

This approach has been implemented in American military activities (see directive DoDD 5000.39 relating to the acquisition and management of integrated logistics support). The practice of ILS is a part of the process of eco-design, because its application begins during design of the product. Yet this approach goes further, because it relates also to production, sales, support and finally end-of-life removal of the product.

The general concepts of integrated logistics support are set out in the AFNOR norm NF X 50 420: 1994. Consequently, integrated logistics support is part of a salient element of GrSCM, as later pointed out by Srivastava [SRI 07].

Thus, Bardou [BAR 00, p. 7] associates six objectives with integrated logistics support:

– "controlling the overall couple 'cost of minimum possession/ maximum operational availability', throughout the whole of the system's life-cycle, from its design to its scrapping;

– taking account of the support requirements in the design of the system (instructions, specifications, etc.) and influencing decisions;

– overall analysis, including the operational system and its support system, necessary and sufficient. Simultaneous examination in order to allow for interactions as early as the feasibility study;

– ensuring consistency of the support elements, amongst themselves (prescribed procedures, using the predefined tools and spares, described in the documentation, used for training of staff, etc.), and with the system;

– adapting to the users' requirements, by way of a policy of constant exchange of information and data with them;

– verifying the aptitude for support, after installation, by a practice of feedback and constant evaluation."

2.1.2. *The components of green supply chain management according to [SRI 07]*

2.1.2.1. *The importance of green supply chain management*

The consideration of the necessity and importance of GrSCM is crucial. It suggests an implementation of "*different approaches to explore the area*" (p. 57). The intention is to advocate a reactive, proactive and value-creating approach. The competitive advantage is sought here.

Hence, the role of the manager is extremely important, as are interactions with the stakeholders. Finally, particular attention must be paid to performance measurement. Thus, the importance of GrSCM highlights the strategic value of this approach.

2.1.2.2. *Green design*

Green design is made up of two sub-elements: environmentally conscious design (ECD) and lifecycle assessment/analysis (LCA). The aim of green design is to "*develop an understanding of how design decisions affect a product's environmental compatibility*" (p. 58). By this factor, one should understand risk management, product safety, workplace health and safety, pollution prevention, resource conservation and waste management.

To begin with, these criteria are in accordance with eco-design. According to the directive 2005/32/EC (European Parliament) "*establishing a framework for the setting of ecodesign requirements for energy-using products*" *(EuP), ecodesign means:* "*the integration of environmental aspects into product design with the aim of improving the environmental performance of the EuP throughout its whole life cycle*[1]".

Then, we find the concept of eco-social design as advanced by the ISO 26000 standard, the aim of which is the quest for eco-social efficiency, (1) providing goods and services which satisfy people's requirements and contribute to the quality of life, whilst (2) gradually reducing environmental damage and negative social impacts throughout the lifecycle of the product. Also, for our purposes, this point can indubitably be linked to the approach of social/societal supply chain management, which is the subject of the next chapter.

2.1.2.3. *Green operations*

Green operations include green manufacturing and remanufacturing, reverse logistics and network design, and waste management.

The key concepts associated with green operations are:

– integration of remanufacturing with the internal operations;

---

1 Source: www2.ademe.fr/servlet/KBaseShow?sort=-1&cid=96&m=3&catid=12922, accessed December 2012.

– comprehension of the effects of competition throughout the process of remanufacturing;

– integration of product design, returned products and bonuses into the supply chain;

– integration of remanufacturing and reverse logistics with the supply chain.

At this stage of Srivastava's summary work [SRI 07], we note a broader description of the elements which make up green operations. Thus:

– green manufacturing and remanufacturing include five stages:

- reduction,

- recycling,

- "remanufacturing", comprising the recovery of the product and/or material and the reuse (repair/renewal and disassembly (with staggering of the disassembly and planning of the process of disassembly)),

- inventory management,

- production, scheduling and planning;

– reverse logistics and design of the network include four points – namely:

- collection,

- inspection/assortment,

- pre-processing,

- location and distribution;

– waste management includes:

- reduction at the source,

- pollution prevention,

- disposal.

2.1.3. *Our proposal for the aspects of the environmental dimension of sustainable supply chain management*

In the context of this book, we modify Srivastava's approach, extracting the social aspect so as to focus only on the environmental aspect. Similarly, we introduce a dimension relating to green transport. Finally, the strategic importance of GrSCM – as shown in Figure 1.1 (sustainable supply chain management in its economic aspect) – becomes a crucially important element in this "thinking". Consequently, the arrangement is as shown in Figure 2.1.

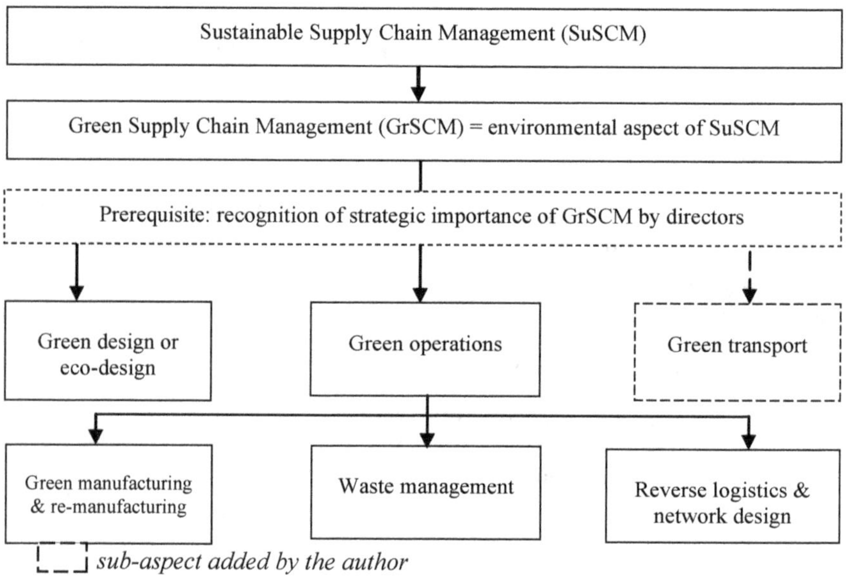

Figure 2.1. *The environmental aspect of sustainable supply chain management*

2.2. Green design or eco-design

One of the first elements to implement for any company wishing to integrate green thinking consists of practicing ecological design. From the moment of research and development, this ecological design, or eco-design, reflects on the launch of a product in accordance with an approach of sustainable production. The main objectives therefore need to be considered. Firstly, we need to identify those methods which use least resources in the process of production; then, we need to design the product with a view to maximum reusability of the different parts at the end of its lifecycle [GHE 05; ADE 06]. Hence, it should be noted here that the approach of eco-design is inspired by and integrates approaches founded on lifecycle analyses (LCAs) of the product [REY 07]. Remember that along with environmentally-conscious design, LCA of the product is one of the two elements which make for "green design", in Srivastava's view [SRI 07].

> ISO/IEC 15288 standard: 2008 – systems and software engineering – System lifecycle processes (source: http://www.iso.org/iso/home/news_index/news_archive/news.htm?refid=Ref855) "[It] offers a portfolio of generic processes for the optimal management of all stages in the life of any product or service, in any sector."

Among the approaches upon which eco-design is founded, Berneman *et al.* [BER 09] cite eco-efficiency. Eco-efficiency was introduced at the Rio Summit by Stephan Schmidheiny. It was then defined in detail by the World Business Council for Sustainable Development. Eco-efficiency seeks to:

– reduce materials intensity;

– reduce energy intensity;

– reduce the dispersion of toxic substances;

– improve recyclability;

– maximize the use of renewable resources;

– prolong products' lifecycles;

– increase service intensity.

2.2.1. *Environmental management system*

Internal advantages	External advantages
Rationalization of production / reduction of costs	Improved competitiveness / competitive advantage
Respect of environmental laws / legal security	Better image in the eyes of the customers and the public
Technological innovations / pollution prevention	Better relations with the authorities / active cooperation
Motivation of collaborators	Greater transparency to shareholders, banks, insurers, etc.

Table 2.1. *The advantages of an environmental management system (source: [SIM 03])*

The first step in an eco-design approach is to put in place an environmental management system (EMS). This system is defined as a "component in an organization's management system used to develop and implement its environmental policy and manage its environmental aspects" (ISO 14050 standard: 2009). Simonet [SIM 03] identifies many internal and external advantages to the implementation of an EMS in an organization (see Table 2.1). Two referential frameworks are associated with the EMS: the ISO 14001 standard and the European system EMAS (Eco Management and Audit Scheme). We shall go into detail about these elements later on in this chapter.

That said, Reyes Carrillo [REY 07] demonstrates four shortcomings of an EMS:

– it may omit the environmental impacts relating to the product [AMM 05];

– the interpretation of the directional lines may be different from one actor to another [ZUT 04];

– it describes the results to be desired, but not the way in which they are to be achieved [MAC 05];

– it is not appropriate for SMEs/SMIs[PER 98].

Hence, Reyes Carrillo [REY 07] stresses that companies need to go beyond an EMS and implement an eco-design approach.

2.2.2. *The eco-design approach*

At the heart of the eco-design approach, a common measurement relates to the recycling of vehicles. Thus, EC directive 2000/53/EC of 18 September 2000 on end-of-life vehicles (ELV) set concrete objectives to be attained by 1 September 2015, which consist of the obligation – in terms of average weight per vehicle and per year – to reuse and draw value from at least 95% of vehicle parts; and to reuse and recycle at least 85% of vehicles. Thus, the automobile sector has to start thinking from the very earliest stage about how to design a "cleaner" product. In this respect, this eco-design approach cannot be envisaged without the support of logisticians, and particularly production logistics managers, to verify the feasibility of a project on the basis of the machinery and skills available.

Although concern about energy issues dates from the 1950s/1960s, it was not until May 1988 that the publication of the booklet FD X 30-310 laid down the bases for eco-design, and the norm XP ISO/TR 14062 of 2002 took account of the environment in design, marking the true beginning of eco-design in industries. AFNOR defines eco-design thus:

> Eco-design or the integration of environmental aspects into the design and development of a product (goods and services) aims to reduce the negative environmental impacts of the products throughout their life-cycle, whilst preserving the quality of use of the product, or indeed improving it.

Eco-design has its origins in the work of Victor Papanek[2], who worked on the notion of eco-design. Many typologies exist in terms of the application of eco-design.

Brezet and Van Hemel [BRE 97] present for stages of eco-design:

– improvement of the product;

– re-design of the product;

– innovation of the product;

– system innovation.

Millet *et al.* [MIL 03], for their part, prefer a three-stage typology which would also correspond to a time-based logic:

– in the 1980s-1990s, partial eco-design reflected an eco-design on part of the life cycle of the product (here we see again the logic of "re-manufacturing" as highlighted by Srivastava [SRI 07];

– between 1990 and 2005, classic eco-design, which focuses, from the very earliest stage of research and development, on an environmentally-sustainable design;

– from 2005 onwards, innovative eco-design, which goes beyond the product and seeks to provide a sustainable service.

Berneman *et al.* [BER 09] give an overview of the economic publications on the use of eco-design and highlight three main categories of economic advantages – namely:

– an increase in revenue, due to the customers' choice of eco-designed products, qualification of suppliers, product differentiation and customer loyalty;

2 Victor Papanek (1923-1998) became a strong advocate of the socially and ecologically responsible design of products. He disapproved of manufactured products that were unsafe, showy, maladapted or essentially useless. His products, writings and lectures were collectively considered an example and spur by many designers. He wrote that "design has become the most powerful tool with which man shapes his tools and environments." (source: http://en.wikipedia.org/wiki/Victor_Papanek).

– a reduction in costs. This point is dealt with in detail in [ADE 99], which points out avenues for reducing costs by way of recyclability of the product and the materials, control of risks of the materials and substances, renewal of the materials, optimization of the materials and the supply chain, energy saving, renewable energies and a strategy of sustainability;

– regarding revenues or costs. For instance, the obligation to collect the products at the end of their lifecycle by a company that practices eco-design proves more competitive. Eco-designal so has a part to play in relationships with the stakeholders, in creativity and finally in enhanced productivity on the part of the employees.

It should be noted that these three categories supplement the elements cited by Simonet [SIM 03] in relation to the Environmental Management System.

The list of European directives on eco-design:

– directive 2000/53/EC on end-of-life vehicles;

– directive 2002/95/EC, or RoHS, for "Restriction of the use of certain Hazardous Substances in electrical and electronic equipment": relates to the management of WEEEs (waste electrical and electronic equipment); followed by Directive 2002/96/EC;

– directive 2004/12/EC on packaging and packaging waste;

– directive 2005/32/EC, or EuOp: on products with high greenhouse gas emissions (heating and hot water production equipment, electrical motor systems, lighting in the residential and tertiary sectors, household apparatus, office equipment in the residential and tertiary sectors, publicly-available electronics, heating systems, ventilation, air conditioning, etc.);

– directive 2006/121/EC, or REACH, for "Registration, Evaluation and Authorization of Chemicals";

– directive 96/29/Euratom on radioactive products;

– directive 2009/31/EC on the geological storage of carbon dioxide prevention and reduction of pollution.

In relation to eco-design, we should also stress the importance of standardization of products by the implementation of an environmental declaration, better known by the term "ecolabel". The European ecolabel was set up by the (EWC) regulation n° 880/92, issued by the Council on 23 March 1992, published in the Official Journal of the European Community on 11 April 1992. At the time of writing, the regulation in force is the (EC) regulation n° 66/2010 of 25 November 2009, applicable from 20 February 2010.

2.3. Green operations

2.3.1. *Green manufacturing and remanufacturing*

For Srivastava [SRI 07], green manufacturing and "remanufacturing" can be characterized by the following five elements:

– reduction,

– recycling,

– "remanufacturing", which involves the recovery of the product/material and its reuse,

– inventory management,

– production, scheduling and planning.

Reduction is echoed in eco-design. Inventory management refers to all the tools and/or thought processes to help maintain an inventory as low as possible and sustainable (no obsolete products, best possible placement depending on the rate of rotation and/or the physical peculiarities of the product). We describe the concept of "remanufacturing" below. As far as recycling is concerned, it is part of the process of waste management and reverse logistics.

Remanufacturing is an interesting technique, in the sense that its aim is to give a used product a new lease of life. The re-valuing involves the integration of new components into aged equipment. In this way, the device in question is able to function more efficiently on the market, performing with equivalent quality to a new product. The environmental advantage to this technique lies particularly in the fact that this process does not require a whole new product to be manufactured, but instead reuses an already existing base [MIC 10]. The approach of remanufacturing is largely used on industrial products, by certain original equipment manufacturers (OEM) [ELK 10].

Chouinard *et al.*[CHO 06] say that two modes of remanufacture are possible. First, there is *remanufacture-to-stock*, which is a push-flow strategy, the aim of which is to immediately deal with the recovered product: the advantage of this is to be able to respond proactively to any subsequent demand. Then, there is *remanufacture-to-order* which, for its part, is a pull-flow strategy, and therefore does not revisit the product until and unless the need is clearly identified.

2.3.2. *Waste management*

Waste management went through a number of stages before the actors involved turned their attention to the logistics of waste. French law 75-633 of 15 July 1975 defines waste as:

> any residue from a process of production, transformation or use, any substance, material, product or more generally any good which has been abandoned or whose owner intends to abandon it.

We can distinguish a number of different categories of waste:

– household and municipal waste: the elimination of such waste is taken care of by collectives;

– industrial waste, which results from a professional activity. This includes:

- non-hazardous industrial waste (NHIW) considered non-dangerous waste. This category includes paper, cardboard, plastic, wood, glass, textiles, rubber, etc.,

- packaging waste, which belongs to the category of NHIW but stems from a specific regulation (e.g. palettes, paper packaging/non-soiled cardboard),

- hazardous waste, including:

– special industrial waste (SIW): these waste products require specific treatments, because they contain toxic and dangerous elements;

– toxic waste in dispersed quantities (TWDQ): similar to SID, but produced on a smaller scale,

- inert waste, which is mineral waste that is not susceptible to physio-chemical changes – e.g. debris or rubble;

– organic waste, from agriculture and agro-food industries such as animal droppings (manure), crop residues (straw) and the residues from manufacture in agro-food industries (blood, bones, etc.);

– end waste products (French law, 13 July 1992): these are waste products from which no more technical or economic value can be gleaned (e.g. office sweepings, mixed household refuse or indeed plastic garbage containers, depending on the choice of the municipal authority).

EXAMPLE.– "Chariots: Jungheinrich s'attaque aux déchets industriels dangereux" (Chariots: Jungheinrich gets stuck into hazardous industrial waste), *Supply Chain Magazine*, 2 March 2012.

> Since the start of this year, the company Jungheinrich has been offering a solution for the management of HIW (Hazardous Industrial Waste). This new organization is intended to respond to both the regulatory requirements and the demand of some of their customers, as Matthias Windwehr, head of SAV France, explains: "The Environmental Code (article L.541, decree N° 2005-665 of 30/05/2005, ministerial decree of 29/07/2005) stipulates that producers of HIW must ensure the traceability and management of these hazardous waste products until they are eliminated, by way of a BSD (*Bordereau de Suivi de Déchets Dangereux* – Hazardous Waste Tracking Document). However, the code does not specify who that producer is: is it the manufacturer, the owner, the user or the maintenance personnel? Thus, in

> the absence of clear and precise regulations, we have chosen to offer our customers a solution which deals with all waste products created by their after-sales activities. With each intervention, technicians are sent to remove all of the HIW. While this arrangement relieves our customers of the logistical concerns relating to the evacuation, processing and traceability of their HIW, it is also to be viewed as a responsible partnership approach."
> The Jungheinrich Group, mainly known in France as a manufacturer of forklift trucks, is also an international actor in the area of storage and management of the flow of goods.

The collection and processing of waste takes place in accordance with rigorous regulations. Each category must respect traceability procedures, which necessitates strict and complicated administrative management. For instance, since 1 December 2005, a Hazardous Waste Tracking Document must be submitted to the Ministry for the Environment for radioactive waste and hazardous waste as defined in the nomenclature of the decree of 18 April 2002 on waste classification. Any company which does not issue this document or issues it incompletely is liable for a fine of up to €750.

2.3.2.1. *The case of electrical and electronic equipment*

There are various categories of waste relating to complex manufactured products: electrical and electronic equipment. We can cite:

– "gray" goods, relating to ICT (computers, office technology, telephony equipment);

– "brown" goods, relating to audiovisual use (televisions, hi-fis, video recorders);

– "white" goods, relating to household electronics (small household electronics, refrigerators);

– industrial goods (electrical cabinets, inverters);

– manufacturing refuse (circuit breakers, contactors).

> Decree n° 2005-829 of 20 July 2005 (Journal Officiel n° 22) specifies what we mean by "electrical and electronic equipment".
>
> Article 1. "This decree applies to electrical and electronic equipment and to the waste products issued from such equipment, including all the components, sub-sets and consumables which form an integral part of the product when it is discarded. Electrical and electronic equipment refers to devices which function because of electrical currents or electromagnetic fields, and devices for generating, transmitting and measuring such currents or fields, designed to be used at a voltage no higher than 1000 volts of alternating current and 1500 volts of direct current… The field of application of this decree does not include:
>
> – electrical and electronic equipment forming part of another type of equipment which, itself, cannot be qualified as electrical or electronic equipment in the sense of this decree;
>
> – electrical and electronic equipment relating to the defense of the essential interests of the State, armaments, munitions and other war materials, if they are for exclusively military ends."

Annex 1 to Decree 2005-829 divides electrical and electronic equipment (EEE) into ten major categories:

– large household goods (refrigerators, freezers, electric cookers, washing machines, tumble dryers, etc.);

– small household goods (coffee-makers, vacuum cleaners, sewing machines, toasters, irons, hairdryers, etc.);

– computing and telecommunications equipment (computers, printers, fax machines);

– consumer equipment (hi-fi systems, radio sets, televisions, camcorders, video recorders);

– lighting material, with the exception of domestic lighting devices and filament bulbs (energy-saving lamps, fluorescent tubes);

– electrical and electronic tools, with the exception of large fixed industrial tools (electric drills, saws, gardening tools);

– toys, leisure and sport equipment (games consoles, remote-controlled toys);

– medical devices, with the exception of any implanted and infected products (blood pressure monitor, electronic weighing scales, thermometers);

– surveillance and monitoring instruments (smoke detectors, video surveillance devices);

– vending machines.

The management of electrical and electronic equipment is subject to multiple restatements. Thus, any EEE which is the object of selective collection must be marked with the symbol of "a wheeled garbage can with a strikethrough cross", visibly, legibly and indelibly. If the dimensions of the equipment are such that this pictogram cannot be displayed, it must be shown on the packaging and on the warranty documents and usage instructions which accompany the product (article 6 of Decree n° 2005-829).

It should be noted that electrical and electronic equipment and Waste electrical and electronic equipment (WEEE) are regulated on a European level by two directives:

– directive 2002/95/EC-ROHS of 27 January 2003 on the limitation of the use of certain hazardous substances in electrical and electronic equipment;

– directive 2002/96/EC-WEEE of 27 January 2003 on waste electrical and electronic equipment.

It should also be noted that since 13 August 2005, large distribution stores are obliged to accept the trade-in of any old device against the purchase of a new one, if the customer desires.

2.3.2.2. *Eco-organizations that deal with WEEE*

For instance, in France, there are four eco-organizations which collect and manage waste electrical and electronic equipment:

– Eco-systèmes,

– Ecologic,

– ERP (European Recycling Platform),

– Récylum.

Table 2.2 gives an overview of these four organizations.

Eco-systèmes	(Source: www.eco-systemes.fr/)
"Set up in July 2005 by 35 producers and distributors, Eco-systèmes is an eco-organization sanctioned by the public authorities, which has been operating since 9 August 2006. In line with European Directive 2002/96-EC and the Decree of 20 July 2005, the mission of Eco-systèmes is one of general interest: to put in place and operate throughout the French national territory, since 15 November 2006, a national system to collect, de-pollute and recycle waste electrical and electronic equipment (WEEE) at end of life, with the exception of lamps. Eco-systèmes is a privately-funded, not-for-profit simplified joint-stock company. The board of directors is headed by Alain Grimm-Hecker (of Groupe SEB France). Christian Brabant is the CEO of the eco-organization, which has 57 employees."	
Ecologic	(Source: www.ecologic-france.com/)
"Ecologic takes care, on behalf of producers, of the end-of-life processing of electrical and electronic equipment (EEE): collection, de-pollution, recycling and value creation. Ecologic was founded in December 2005 on the initiative of thirty EEE producers, supported by the professional federations FICIME and ALLIANCE TICS. Ecologic is the first eco-organization to receive both ISO 9001 and ISO 14001 certification."	

Table 2.2. *The eco-organizations which deal with WEEE*

<table>
<tr><td>ERP</td><td>(Source: www.erp-recycling.org/#)</td></tr>
<tr><td colspan="2">"The European Recycling Platform is an eco-organization set up in 2002 to enable its members to conform with the new European regulation regarding Waste Electrical and Electronic Equipment (WEEE).

This eco-organization is a structure which takes care of the removal and processing of WEEE on behalf of its members. It is managed in such a way that enables it, for instance, to choose to outsource to external contractors for the operations of collection and processing.

It was set up on the initiative of four European producers – namely Braun, Electrolux, Sony and HP.

It is managed by Geodis Valenda, which offers collection, logistical services, recycling, management operations, reporting and cost optimization."</td></tr>
<tr><td>Récylum</td><td>(Source: www.recylum.com/)</td></tr>
<tr><td colspan="2">"Official eco-organization for the collection and recycling of used lamps.

Set up on 26 May 2005 by a number of lamp manufacturers wishing to intensify their efforts in terms of environmental protection, Récylum takes action on behalf of all lamp producers who so desire (manufacturers, importers, distributors with their own brand, etc.).

On 19/07/2010, 614 companies chose to use Récylum to fulfill their legal obligations as producers."</td></tr>
</table>

Table 2.2. (continued) *The eco-organizations which deal with WEEE*

2.3.3. *Reverse logistics*

"Reverse distribution" was one of the earliest terms to be used in reference to reverse logistics. Carter and Ellram [CAR 98] paint it as "the return, upstream movement of a good or material resulting from reuse, recycling, or disposal. This upstream movement can be associated with environmental as well as quality and wear-dating issues, and it is often performed by new, auxiliary channel members."

In the eyes of certain authors, green logistics also has a part to play in a reverse logistics approach (see Table 2.3).

Authors	Definition
Rogers and Tibben-Lembke [ROG 01]	Green logistics consists of making "efforts to measure and minimize the environmental impact of logistics activities."
Murphy and Poist [MUR03]	Green logistics has a bearing on logistical systems in terms of thirteen points: transport, warehousing, manipulation of materials, packaging, inventory management, the ordering process, purchasing, demand prediction, building implantation, customer service, production planning, recovery and destruction of waste, handling of product returns and international logistics.

Table 2.3. *The definitions attached to green logistics*

Yet above all, it is the term "reverse logistics" which predominates in current discourse (see Table 2.4).

Authors	Definition
Beaulieu *et al.* [BEA 99a]	This can be viewed through the lens of two types of returns. Firstly, recycling of products; then, returns or recalls of products to the manufacturers (e.g. faulty or contaminated products). Reverse logistics includes four elements: – reverse logistics includes both a product and a waste product with a value in recovery or reuse; – the point of return of these products or waste products in are verse logistics network may be the initial producers, a different player from the same industry or an actor from other sectors of activity; – reverse logistics also involves more than the simple choice of distribution network; – reverse logistics also includes a set of management activities. It is "a set of management activities aimed at reintroducing secondary assets into added-value chains."

Table 2.4. *The definitions attached to reverse logistics*

Rogers and Tibben-Lembke [ROG 99]	The process of planning, implementation and control, rationally and advantageously, of the flows of raw materials, works-in-progress, finished products and information relating to such products, from the point of consumption to the point of origin, with the aim of recovering or creating value or improving the disposal of waste.
Lambert and Riopel [LAM 03]	The process of planning, implementation and control of the efficiency and profitability of raw materials, works-in-progress, finished products and pertinent information from the point of use to the point of origin, with the aim of recuperating or generating value or to correctly possess it, whilst ensuring effective and environmentally viable usage of the resources brought to bear.

Table 2.4. (continued) *The definitions attached to reverse logistics*

Nevertheless, in the eyes of Rogers and Tibben-Lembke [ROG 01], a distinction has to be drawn between green logistics and reverse logistics, particularly in terms of the handling of the products. Green logistics touches specifically on eleven areas:

– energy conservation;

– material conservation;

– efficient use of lands;

– reduction of traffic;

– reduction of air pollution;

– reduction of water pollution;

– reduction of visual pollution;

– odor reduction;

– noise reduction;

– management of solid waste;

– management of hazardous waste.

Thus, reverse logistics relates to the return of products, unsold goods and de-stockage onto secondary markets.

We hear a rather different story from Lambert and Riopel [LAM 03], for whom reverse logistics combines reverse distribution and green logistics (see Figure 2.2).

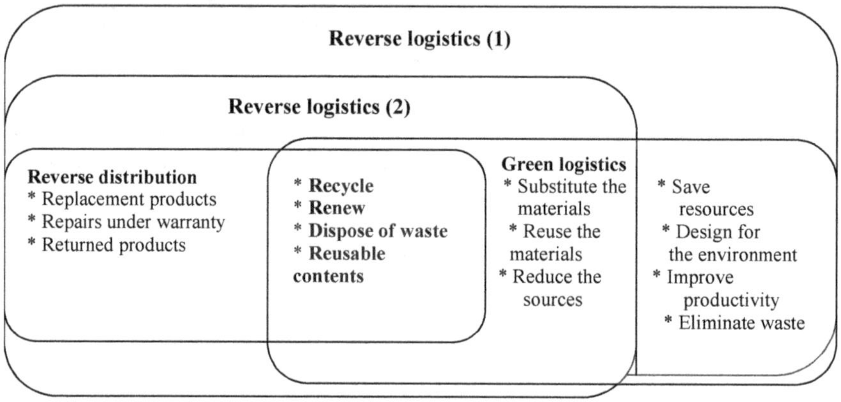

(1) Definition advanced by Lambert and Riopel [LAM 03];
(2) Definition advanced by Rogers and Tibben-Lembke [ROG 98]

Figure 2.2. *Reverse logistics*

Yet in any case, the primary aim is to use as little as possible, or to reuse as fully as possible, because of the omnipresent issue of the rarefaction of the raw materials.

2.3.3.1. *The ecological, strategic and financial roles involved in reverse logistics*

In general, the benefits to be expected from reverse logistics are of three types: ecological, strategic (e.g. the positive impact for the clientele of products based on recycled materials) and financial (because of proactive management or better knowledge of shortcomings).

Beware, however: if mishandled, reverse logistics may cause a loss of customer confidence, the symptoms of which can be diagnosed when:

– returns arrive in a shorter time-span than the projected operational lifespan of the product;

– there is the presence of a great deal of returned stock in the warehouse;

– there is the presence of unidentified or unauthorized returns;

– the cycle time taken to deal with returns is overly long;

– the cost of processing a return is unknown[DAW 95].

To begin with, the ecological role of reverse logistics is plainly apparent (less pollution, reassignment).

Next, the strategic role of returns can be appreciated through the lens of six reasons [ROG 98] which are:

– competitive reasons, which refer to liberal returns policies as practiced in the United States, whereby any consumer can return a product, defective or otherwise; and to altruistic reasons such as the return of products in compensation of partial purchasing vouchers and the reuse of these products in the context of humanitarian causes;

– the proper channel, which is based on a logic of recycling of the old for future resale;

– the regulation, which favors a raise in taxes on products which cannot be reused, and which therefore require a better *a priori* management of the manufacture of the products;

– value recovery;

– the recovery of assets which shed light on all the advantages – in terms of image and financial viability – of efficient management of reverse flows;

– margin protection, which highlights the positive effect of maximization of inventory management, e.g. by decreasing the risk of obsolescence.

Finally, we note that the financial aspect constitutes a crucial element in terms of the decision to adopt a reverse logistics program. As De Brito points out [DEB 04], direct and indirect financial gains are made from reverse logistics: direct financial gains when it is possible to re-valorize a product with high economic value, and indirect financial gains in the context of anticipation (here we see once more the aspect linked to remanufacturing).

2.3.3.2. *The activities attached to reverse logistics*

In 2004, Lambert and Riopel put forward an integrative model of reverse logistics which pays equal attention to the retailer and to the company. Here, only the return of products is considered, leaving aside the issue of the flow of documents required by this type of activities. They base their thinking on the four main stages – the barrier (entry gate), collection, sorting and choice of processing (disposal) [ROG 98], with which they associate a certain number of activities.

2.3.3.2.1. Stage 1: the barrier

The barrier comprises "two" successive activities: "the need for a return" which leads to the "barrier".

In regard to the need for a return, three major categories of justifications are distinguished: commercial reasons, legal responsibilities and financial reasons:

– commercial reasons have the primary objective of enhancing the customer's level of satisfaction. Two sub-types are included in this first justification:

- technical problems (or after-sales service) including repair (under warranty or otherwise), renewal and a recall campaign:

– the company seeks to implement "technical assistance" solutions (by telephone, over the Internet, in store) before approving a process of return of the product. This approach of "technical assistance" constitutes a non-negligible source of information for the improvement of the products and services provided;

– as regards the aspect "recall campaign", more specific to the first two aspects, the company is constrained to maximum customer information and contact,

- commercial recalls relating to excess stock or a dispatch error. If *ex ante*, a significant quantity of stock is put in place to cater for an increase in sales, a return will result from poor inventory management (overproduction). A dispatch error, for its part, is relatively uncommon,

– legal responsibilities refer to the increase in legislation which imposes a decrease in the quantity of waste buried. Here we are speaking of end-of-life recycling of products, packaging material or indeed production refuse. Nevertheless, at this stage, the company may itself decide to lean towards an environmental philosophy, for instance, by obtaining ISO 14000 certification. It should be noted that for certain products, the barrier stage may be bypassed and we may go directly to collection (e.g. laser printer cartridges, reusable containers);

– economic reasons relate to reuse (packaging material, reusable containers, hired products) and recovery (production waste). In particular, these reasons highlight the potential savings to be made, rather than the supposed environmental gains. The choice is made when the return represents a greater financial benefit than the manufacture of a new product or the purchase of new raw material.

The need for a return constitutes an important stage and plays the role of a filter before the company becomes involved. It verifies that the return is indeed – contractually or legally – a matter of concern for the company.

Remember that in certain cases, this stage can be circumvented. Indeed, when we know from the very outset of the purchasing process that the product can be reused, revalorized or recycled at the end of its life, this initial control layer is not necessary (e.g. ink cartridges, batteries, etc.).

As regards the second activity – the barrier – this verifies the well-foundedness of the return. Three scenarios may present themselves:

– the returned product is associated with a replacement product;

– the product is accepted by the company for examination and appropriate treatment in the case of the return in question;

– the company refuses the return for contractual reasons or because of lack of information, but the customer decides to return the product nevertheless. In this last scenario, the company has to have an "exceptional case mechanism" put in place to deal with the return.

2.3.3.2.2. Stage 2: collection

Collection comprises two activities: "removal of the product" and "transport of the returns".

As regards the removal of the product, there are many ways in which it can be removed. The customer can take it back to the point of sale, or to an authorized center. It can also be sent via the postal service or assimilated. Finally, an authorized person (the selling company or a contractor working under the company's authority) can come to the customer's home directly to pick it up. However, it may occur – for certain products such as electronic goods – that a service technician comes out in order to assess what action(s) need(s) to be undertaken. Thus, it is possible to carry out a repair on the spot, replace the product or indeed take away the defective product.

The second activity of collection relates to the transport of the returns. The choice of destination for the returned good depends – in no small part – on the status and/or the state of the said product when it is returned. Thus, it may be immediately recycled. When the product reaches its destination, stage 3 (sorting) begins.

2.3.3.2.3. Stage 3: sorting

Sorting enables us to establish a diagnosis for each product and make the appropriate decisions on that basis (dispatch of a replacement product, reimbursement, and repair). The more complex the network, the more unwieldy the task of management of stocks in transit and transport between the different sites, making this an important activity.

2.3.3.2.4. Stage 4: processing

Processing involves sending the product to its final destination. Three actions are associated with this stage:

– a detailed inspection of the returned product: this activity complements the preliminary choice of processing made during the sorting stage. The aim is to decide on the final remedial treatment to be undertaken;

– stock management: the goal is “to confirm that the product can enter into the process of treatment chosen”;

– the treatment itself: eleven possible operations – namely:

- repackaging,
- repairing,
- reusing,
- reconfiguring,
- renewing,
- updating,
- recycling,
- donating,
- revalorizing,
- selling on other markets,
- scrapping.

Here, Lambert and Riopel [LAM 04] specify two main activities at the level of this stage: “inventory management” and “choice of remedial action”:

– inventory management is of crucial importance in any logistical activity. Its impact on the system is at once financial, physical (use of space) and productive (workload management). When we speak of reverse logistics, two elements need to be considered: first, everything relating to the traceability of the product (easier pinpointing); then the existence or not of a “buffer stock”[3] which – all other things being

3 Buffer stock corresponds to the minimum stock needing to be kept in the company to cover the delay of resupplying. It is added to the safety stock (limitation of stock

equal – facilitates an improved response time and better management of the handling of the return;

– the choice of remedial action gives us eleven possible operations:

- repackaging: repackaging is a relatively common operation in the context of commercial returns (e.g. duplicate Christmas presents). The primary objective is to verify the quality of the packaging, so as to be able to put the product back on sale as soon as possible. It goes without saying that the inspection of this return "invites" in-depth examination of the functionality of the product before it is put back on the shelf,

- repairing: before any repair of a material, a diagnosis has to be made. Its aim: to verify whether (1) the product can indeed be repaired and (2) it makes financial sense to repair it. If this is not the case, it is the "recycling process" which is begun. If it can be repaired, a number of choices may manifest themselves: the repair is completely successful and the product is returned to stock if there has been a system or previous compensation, or returned to the customer; the repair is not successful or does not correspond to the quality criteria deemed sufficient on a commercial level and the product either goes to a secondary market or is recycled,

- renewing: renewing shadows the process of repairing quite closely. The difference is to be appreciated in the following point: "renewing changes more than simply the defective parts or components of the product; it changes a series of parts and components so as to restore the product to a new or nearly-new state",

- updating: this process is similar to the previous process (renewing), with the sole difference that the aim is not necessarily to prolong the life of the product, but rather to imbue it with a financial added value on resale. In fact, it is the improvement of the function or appearance of the product which is mainly sought here,

---

shortfalls in the face of various risks, such as inaccurate predictions to give alert stock or critical stock, which is the moment to issue an order so as to avoid any shortfall).

- reusing: his is synonymous with recovering when we speak of reusing parts from a product. The objective of recovery is to "reduce the costs and environmental impact by using a good more than once",

- reconfiguring the product: the objective of this treatment is to change a certain number of parts in order to cater for the customer's requirement (e.g. reconfiguration of a car for a new market),

- recycling: the goal of recycling is to reduce exhaustion of resources: "This process involves dismantling the products and separating the materials (ferrous and non-ferrous metals, plastic, glass, paper, etc.). Each of the separate materials is put in appropriate containers for resale or later reuse. The finer the separation of materials, the more the value will increase",

- donating: if the good is still functional, it is donated to not-for-profit organizations such as charities,

- revalorizing: this consists of reusing the good within the company itself (e.g. the oils used). The good can also be resold,

- selling on other markets: in general, the good will be resold on a secondary market because it is of inferior quality,

- scrapping: the product is sent for landfill or for incineration.

2.3.3.3. *Choice of path and extent of treatment*

Beaulieu *et al.* [BEA 99a] propose to classify reverse logistics networks according to two parameters:

– the choice of the added-value trajectory along which the "secondary asset" will be directed;

– the extent of processing of that secondary asset (intermediary processing or reprocessing) before it is reintroduced into circulation. This parameter is aimed at giving us an appreciation of the number of determinants involved in the network.

For this purpose, they put forward a typology whose aim is to enable managers to locate the secondary asset and, in fact, evaluate *ex ante* the efforts needing to be made in terms of the implementation of a reverse logistics network (see Figure 2.3).

To explain this diagram, the authors make the following observations: "For example, the palettes used to transport goods and returned products (bought from a catalog) are examples of secondary assets which do not require any transformation and are fed back into their original industry because they will be reused by the distributors and retailers (top left-hand corner).

In the same quadrant, we find recycled soft drinks bottles which only require a minor degree of treatment (i.e. washing). The situation is not exactly the same for old papers. Although they do return to their original industry, the recovered paper has to be subjected to extensive processing to turn it back into a paste, which can then be used for the production of new products (cardboard or paper) (bottom left-hand corner). The products or materials (secondary assets) on the right-hand side of the matrix are more difficult to introduce back into their original industry. There are two explanations for this situation: when these assets are re-treated, it is not possible to return them to the same standards of quality as the new materials, or the secondary asset no longer has any value for the initial producer[…](bottom right-hand corner) […].

A different situation is experienced in the sector of retail trade. In spite of sales, clothes retailers for instance may find themselves with clothes which no longer have any interest for them, because they no longer correspond to the criteria in fashion. Hence, they are donated to charity organizations […] (top right-hand corner)".

The second advantage to this classification is that it highlights the complexity in the management of reverse logistics networks, in two points: (1) "in the difficulty of setting up the network in question, (2) in the difficulty of managing the various interactions between the actors involved in that network."

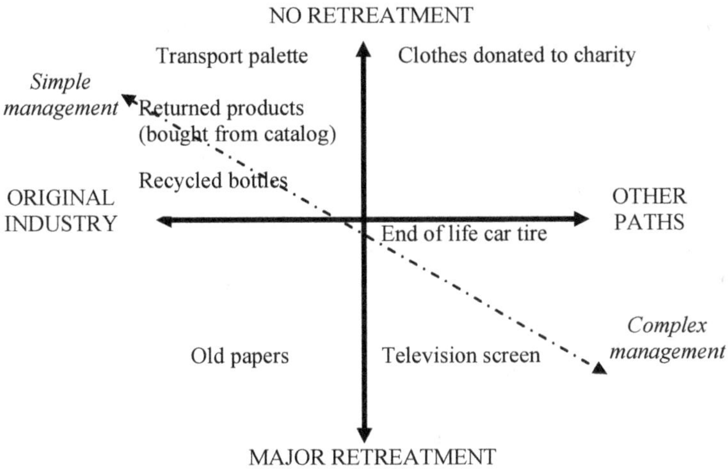

Figure 2.3. *Classification of reverse logistics networks (source: [BEA 99b])*

Beaulieu [BEA 00] supplements this analysis with a detailed description of each of the parameters of the graph previously put forward by Martin, Landry and himself [BEA 99b]. Thus:

– on the reprocessing axis, he envisages three situations:

- summary reprocessing: "the secondary asset retains its physical characteristics and its functionalities are intact. Only operations of conditioning or cleaning are necessary",

- major reprocessing: "the secondary asset retains its physical characteristics but its functionalities are degraded. Repairs are necessary – e.g. the replacement of defective or exhausted parts",

- radical reprocessing: "the user wishes to restore the basic properties. Thus, the asset may be ground up, dismantled, melted down, de-inked, etc.";

– on the distribution network axis, three situations are also suggested:

- use of the current network: "an organization can entirely make use of its current distribution network",

- use of new intermediaries: "an organization can partly make use of its current network but has to involve other actors for the activities of reverse logistics",

- a different path: "in this situation, the secondary asset is sent to a sector of activity other than that which initially produced it".

Consequently, Beaulieu [BEA 00] identifies four measures which could contribute to a reduction in the complexity of the networks:

– from the phase of design of the product, initiatives can be taken to reduce the number of components, and there by *a posteriori* avoid the proliferation of paths of valorization. In this thought process, a high point is to be considered as regards the use of materials the least harmful to the environment possible. Here we hear echoes of the logic of eco-design;

– finding solutions upstream in order to avoid products being returned. For instance, one could implement "precision checks" on the purchasing of products from a catalog or electronically; or design products which are able to integrate improvements from technological developments, which avoids having to replace the product in its entirety. The typical example is a computer;

– with a view to reaching a "critical quantity of assets", the solution is to put in place "incentives" for these assets to be reintroduced into the network. These incentives may take the form of financial penalties (for unreturned products), installations favoring the participation of the users (collection bins) or simplified procedures for returning the products;

– the presence of specialized intermediaries who (1) take care of the distribution of materials when the asset is made up of several components and (2) stand in for the organizations who do not wish to complexify their decisions on stock management by combining a conventional flow and a reverse logistics flow.

2.4. Green transport

Green transport is characterized by its multimodality, i.e. the ability for a company to use several different modes of transport in order to reduce environmental pollution. In that context, companies are encouraged to favor sea, river or rail transport over road and/or air, which are considered heavier polluters.

Another term is part of multimodality – co-modality. Introduced in 2006 by the European Commission, co-modality "positions transport within a global logistical chain" [BLA 08] (p. 14) because "logistics enables us to improve the efficiency of the different modes of transport and their combinations. In addition, more goods should be transported by way of a reduced number of transport units, such as cars, trucks and ships" (European Commission, 2006).

EXAMPLE.– The case of the sea highway: the opening on 16 September 2010 of the sea highway between Montoir-de-Bretagne (Nantes Saint-Nazaire, France) and Gijon (Spain) is part of a multimodality approach. The aim is to relieve congestion on the trans-Pyrenean road networks... provided that traffic is sufficient to ensure the economic viability of the project.

Within the context of green transport we also find last-mile logistics, also known as urban logistics. Last-mile logistics is the subject of a great deal of attention on the ecological scale. Thus, with varying degrees of success, local authorities are putting policies in place to make intercity transport less pollutant and less noisy. In doing so, they are defining limited-traffic zones where only residents and certain commercial vehicles considered to be non-pollutant can go (e.g. hybrid vehicles, electric vehicles, LPG vehicles) [MOR 10].

Green transport fits in with the use of less highly pollutant means of transport outside of cities by way of new generations of heavy goods vehicles[4] and eco-driving policies.

---

4 We identify the generation of pollutant vehicles by way of the so-called "Euro standards" or European emission standards. The first standard, Euro 0, was published

Declaration of CO_2 emissions – Decree n° 2011-1336 of 24 October 2011 on information about the quantity of carbon dioxide emitted during the provision of a transport service.

In the wake of the Environmental Grenelle of 2007, transport operators must declare the CO_2 impact of the service provided, at the bottom of each invoice. This regulatory device will come into force between 1 and 31 December 2013.

> EXAMPLE.– Dossier: "Transport et développement durable : la chasse aux émissions de CO2" (Transport and sustainable development: chasing down CO_2 emissions), "Charte Objectif CO2 : le transport routier s'engage"(CO_2 Objective Charter: road transport joins the fight), *Transports & Logistique*, March 2012, p. 9.
>
> "On 1 January 2012, 484 companies were signatories to the *Charte Objectif CO2*in road transport; 48 of these companies had completed their first three-year cycle of commitment. Spearheaded by MEDDTL and ADEME in collaboration with the professional organizations FNTR, TLF, OTRE and UNOSTRA, the charter was launched in December 2008. Since then, nearly 68,800 drivers and 62,200 vehicles have been or currently are involved in the approach, leading to a reduction in CO_2 emissions and a saving of 144.3 million liters of diesel. According to ADEME, France's fleet of vehicles over 3.5T consumes 10 billion liters of diesel every year. With this in mind, the *Objectif* CO_2 charters have been able to save 1.4%of fuel in relation to overall consumption."

In the context of green transport, finally, we must cite the eco-road tax for HGVs. This tax levied for any usage of the national, local or communal network beyond the agreed quota (around 15,000km). This tax is due to come into force on 1 October 2013 throughout France,

---

in 1990. Currently, it is Euro 5 which is in force (and has been since 2009). Note that Euro 6 will come into force on 1 January 2014 to regulate new industrial vehicles, and has been in force since 1 January 2013 for new models of vehicles.

following an initial test period running from July. The charge will be levied on all vehicles of 3.5 tons or more using France's road network. The law allows for this tax to be passed on in its entirety to the beneficiary of the service, so as not to simply penalize France's road transport network. In itself, it should be noted that this tax will be payable by all vehicles registered both in France and abroad. For instance, one might note the "HGV eco-tax" paid by German HGVs since 1 August 2012, when they take four-lane national roads.

2.5. Systems, regulations, standards and referential frameworks

There are many regulations, standards and referential frameworks associated with an environmentally-conscious way of thinking, and which, by extension, have ramifications for logistics. Below, we present the most widely known of these. Even if these standards concern the whole of an organization, a logistics manager needs to be aware of them, if only in terms of waste management, and particularly waste electrical and electronic equipment, which is part of the practice of a policy of reverse logistics.

2.5.1. *The ISO 14000 standard*

There are a number of standards which make up the ISO 14000 certification on environmental management. The first two are the ISO 14001 and ISO 14004 standards. The ISO 14001 standard of 2004 defines the requirements relating to an environmental management system (EMS). The ISO 14004 standard (2004), for its part, lays out general guidelines for the practice of an EMS. As with any standard, the obtaining of ISO 14000 status results from certification awarded by an organization accredited to award it. According to the ISO itself:

> *At the end of December 2011, at least 267,457 ISO 14001 certification had been awarded in 158 countries, which represents two additional countries and 15,909 additional certifications (+6%) in relation to the previous year.*

China, Japan and Italy score highest for the total number of certificates in this domain; the leading three countries in terms of progression of ISO 14001 certification in 2011 are China, Italy and France. (source: http://www.iso.org/iso/fr/news.htm?refid =Ref1686)

The list of standards in the ISO 14000 family (source: the ISO 14000 family of international standards on environmental management: www.iso.org/iso/fr/theiso14000family_2009.pdf) as detailed below.

Planning:	
ISO 14001:2004	Environmental management systems – Requirements with guidance for use
ISO 14004:2004	Environmental management systems – General guidelines on principles, systems and support techniques
ISO/DIS 14005	Environmental management systems – Guidelines for the phased implementation of an environmental management system, including the use of environmental performance evaluation
ISO/CD 14006	Environmental management systems – Guidelines for incorporating eco-design
ISO 14050:2009	Environmental management – Vocabulary
ISO/TR 14062:2002	Environmental management – Integrating environmental aspects into product design and development
ISO Guide 64:2008	Guide for addressing environmental issues in product standards

Executing:	
ISO 14040:2006	Environmental management – Lifecycle assessment – Principles and framework
ISO 14044:2006	Environmental management – Lifecycle assessment– Requirements and guidelines

ISO/TR 14047:2003	Environmental management – Lifecycle impact assessment – Examples of application of ISO 14042 *(ISO 14042: 2000 – Environmental management – lifecycle assessment – Lifecycle impact assessment)*
ISO/TS 14048:2002	Environmental management – Lifecycle assessment – Data documentation format
ISO/TR 14049:2000	Environmental management – Lifecycle assessment – Examples of application of ISO 14041 to goal and scope definition and inventory analysis *(ISO 14041: 1998 – Environmental management – Lifecycle assessment – Goal and scope definition and inventory analysis)*
ISO/CD 14051	Environmental management – Material flow cost accounting – General framework
ISO/WD 14045	Eco-efficiency assessment – Principles and requirements
ISO 14064-1:2006	Greenhouse gases – Part 1: Specification with guidance at the organization level for quantification and reporting of greenhouse gas emissions and removals
ISO 14064-2:2006	Greenhouse gases – Part 2: Specification with guidance at the project level for quantification, monitoring and reporting of greenhouse gas emission reductions or removal enhancements
ISO/WD 14067-1	Carbon footprint of products – Part 1: Quantification
ISO/WD 14067-2	Carbon footprint of products – Part 2: Communication
ISO/AWI 14069	GHG – Quantification and reporting of GHG emissions for organizations (carbon footprint of organization) – Guidance for the application of ISO 14064-1

Checking:	
ISO 14015:2001	Environmental management – Environmental assessment of sites and organizations (EASO)
ISO 14031:1999	Environmental management – Environmental performance evaluation – Guidelines
ISO 14064-3:2006	Greenhouse gases – Part 3: Specification with guidance for the validation and verification of greenhouse gas assertions

ISO 14065:2007	Greenhouse gases – Requirements for greenhouse gas validation and verification bodies for use in accreditation or other forms of recognition
ISO/CD 14066	Competency requirements for greenhouse gas validators and verifiers document
ISO 19011:2002	Guidelines for quality and/or environmental management systems auditing

Acting:	
ISO 14020:2000	Environmental labels and declarations – General principles
ISO 14021:1999	Environmental labels and declarations – Self-declared environmental claims (Type II environmental labeling)
ISO 14024:1999	Environmental labels and declarations – Type I environmental labeling – Principles and procedures
ISO 14025:2006	Environmental labels and declarations – Type III environmental labeling – Principles and procedures
ISO/AWI 14033	Environmental management – Quantitative environmental information – Guidelines and examples
ISO 14063:2006	Environmental management – Environmental communication – Guidelines and examples

The ISO 14001 standard constitutes the basic referential framework for ISO 14000 certification. The first version of ISO 14001 was released in 1996, but the version currently adhered to dates from 2004. It sets out the generic requirements for an environmental management system. Its implementation stems from the application of numerous requirements, divided into six chapters:[5]

– Chapter 1: general requirements;

– Chapter 2: environmental policy (commitment of the management team);

– Chapter 3: planning, which relates to:

5 www.prorecyclage.com/reglementation/normes.html

- environmental aspects (atmospheric emissions, noise, odors, etc.),

- legal requirements and other requirements,

- objectives and targets (quantification),

- the environmental management program (responsibilities, means, delays);

– Chapter 4: implementation and operation, including:

- structure and responsibility (who does what?),

- training, awareness-raising and competency,

- communication,

- documentation of the environmental management system,

- control of the documentation,

- control of activities,

- prevention of emergencies and the capacity to react;

– Chapter 5: control and corrective action, with:

- monitoring and measuring,

- non-conformity, corrective action and preventive action,

- registration,

- auditing of the environmental management system;

– Chapter 6: management review.

Although it is unwieldy and bureaucratic to implement (as is any norm), Boiral [BOI 04] notes a number of points which explain why ISO 14001 is recognized by companies. Firstly, like ISO 9000, it is an object of commercial promotion and advertising, as well as a selection of suppliers. Similarly, it improves the company's image and demonstrates its environmental commitment. In fact, while ISO 14001 is a constraint with which an ever-increasing number of companies have to comply, it is nevertheless a way of lending "legitimacy" to the organization, as it demonstrates the performance of the structure in a standardized environmental system.

EXAMPLE.–"Politiques de qualité en environnementale: le cas de SOTRADEL" (Environmental quality policies: the case of SOTRADEL), http://sotradel.com/m-28-the-company.html, accessed June 2012.

– All of SOTRADEL's sites and activities are ISO 9001-certified, and all of its warehouses are in the process of ISO 14001 certification (one site has already been certified).

– The transport companies in the SOTRADEL Group have made a commitment to reduce their CO_2 emissions: they are signatories to the CO_2 Charter with ADEME and la DREAL.

– SOTRADEL is committed to an ISO 26000 corporate social responsibility scheme.

– ADEME accepted SOTRADEL's bid for the "National waste valorization plan" in April 2011: SOTRADEL is the only logistics provider of the 50 winners.

– SOTRADEL was nominated for the "Prix Entreprises et Environnement" (Companies and the Environment Prize) in November 2011, awarded by the *Ministère du Développement Durable* (Sustainable Development Ministry) and ADEME, for its global sustainable supply chain management approach, confirming its view of itself as an eco-citizen company. SOTRADEL is the only transport/logistics provider to have been nominated."

2.5.2. *The Eco Management and Audit Scheme*

The EMAS (standing for eco management and audit scheme), or eco-audit, is a European scheme set up in 1995 by the European Commission. It fits in with an environmental management system (EMS). The EMAS regulation – just like ISO 14001 – fits into the adoption of an environmental policy. It was revised in 2009 (Regulation n° 1221/2009 of 25 November 2009 – Official Journal of the European Union published 22/12/2009) and came into force on 11

January 2010. It repeals Regulation n° 761/2001, which had hitherto governed the system, as well as Commission Directives 2001/681/EC and 2006/193/EC. EMAS registration has to be renewed every three years (or four years, for small organizations).

EMAS regulation helps to evaluate, improve and give an account of the company's environmental performances. Any company which is already ISO 14001-certified is given EMAS certification if it publishes an environmental declaration in line with the criteria prescribed by EMAS.

> First article – Objective of EC Regulation n° 1221/2009 of the European Parliament and of the Council[6]:
>
> "EMAS, which is an important instrument in Action Plan Sustainable Consumption and Production (APSCP) and for sustainable industrial policy (SIP), is intended to encourage constant improvement of the results obtained by organizations in terms of the environment, by the establishment and implementation by these organizations of environmental management systems, systematic, objective and periodic operation of these systems, provision of information about the environmental results obtained and consultation with the public and other interested parties, and by means of active participation of the employees of the organizations and appropriate training."

2.5.3. *Approaches associated with quality of habitat*

While eco-design reflects primarily on the ecological design of a product, it also has a part to play in the construction of a building, including the buildings directly connected with logistical operations. For instance, according to the magazine *Stratégie Logistique* (May 2009, p. 10), the implementation of an HQE® (standing for *Haute Qualité Environnementale* – High Environmental Quality) approach in

---

6 http://eurex.europa.eu/LexUriServ/LexUriServ.do?uri=OJ:L:2009:342:0001:0045:FR:PDF

a logistics network ranges from around 11,800.00 Euros excl. tax for buildings of less than 5000m² to 46,000.00 Euros excl. tax for buildings of over 50,000m².

EXAMPLE.– "L'entrepôt durable fait école" (Sustainable Warehouses are Gaining Ground), *Logistiques Magazine*, September 2009, n° 242, p. 76-83 (summary).

Building a warehouse in line with an HQE® approach means, for example, concrete flooring with fewer joints and therefore less noise, less back problems for drivers, easier maintenance and a longer lifespan. At an internal level, it involves designing a building which combines well-insulated walls, abundant natural light, low-consumption lighting, two-way or natural ventilation, hot water provided by solar panels mounted on the roof, Glulam[7] structural framework, etc. At an external level, it involves collecting rainwater to water local agricultural plantations, positioning the parking lots for light and heavy vehicles so that they do not interfere with one another, etc.

EXAMPLE.– "Stef-TFE, 1ère certification HQE"(Stef-TFE – first HQE certification), *Stratégies Logistique*, 14 April 2011.

"Certivea has awarded Stef-TFE the first ever certification *NF Bâtiments Tertiaires – Démarche HQE – plateformes logistiques/entrepôts frigorifiques* (French norm on tertiary buildings, HQE approach – logistical platforms/cooled warehouses) for the site Stef Les Essarts."

---

7 Glued Laminated Timber (Glulam): Glulam is used for roofing frameworks. It is made of laminated wood glued together with a resin. It is stronger than raw wood, which is an advantage at many levels: (a) fire resistance the same as concrete, (b) better ecological factor than concrete, (c) simplified installation by a single company, whereas a concrete structure requires two companies – one for the main structure and another to build the metal skeleton, (d) lighter than concrete, which makes the foundations lighter, (e) it can span greater lengths than can concrete beams, which facilitates the implementation of lighting, etc. (source: "Le lamellé-collé, nouvel atout environnemental" (Glulam – a new environmental asset), *Logistiques Magazine*, April 2009, n° 238, p. 87-89.

A number of bodies are associated with habitat quality (ADEME, AFNOR, CEQUAMI, etc.). Below we briefly present two of these: the association Qualitel and the HQE association.

The independent not-for-profit Qualitel[8] was set up in 1974 with the aim of informing, certifying, evaluating and monitoring habitat quality. It is also a member of the HQE association, whose goal is to develop the environmental quality of buildings, and of the association Effinergie[9], which encourages the low energy-consumption construction, both for new projects and for renovation, and is developing a set of guidelines on energy performances of buildings.

The French HQE association, founded in 1996, is there to help companies implement the HQE® approach.[10] It was recognized as a public utility by decree on 5 January 2004. In 2012, it had 98 members, who were divided into seven "schools":

– users (two members);

– contracting authorities (27 members);

– advice and support (14 members);

– companies and industries (26 members);

– experts (nine members);

– project managers (seven members);

– honorary members (13 members).

The HQE® approach offers a guide which helps to integrate environmental and sanitary aspects into the different phases of the lifecycle of a building – namely design, construction, lifetime but also

8 www.qualitel.org/accueil/.

9 The association Effinergie was founded on 27 March 2006 with the aim of developing a guiding framework for energy performance. It was in 2007 that the label BBC-Effinergie® for new buildings was published, with its counterpart for renovation projects being released in 2008. In collaboration with Certivéa, the Effinergie® label is being promulgated throughout buildings in the tertiary sector, with or without the HQE® approach.

10 www.assohqe.org/.

deconstruction. Since 2010, the evaluative frameworks impose two elements:[11]

– the implementation of a management system based on ISO 9000 and ISO 9001. It also takes account of the GA AFNOR P 01 030 approach, which appears as a perimeter of application for the EMS framework;

– the declaration of the "environmental quality of the building" (EQB) or its environmental and sanitary performance which, in terms of communication, must:

- describe the operation (usage, location, type, surface, etc.) and is "characterized in terms of technical performances fixed by the agreed specifications (accessibility, thermal and acoustical comfort, mechanical strength, seismic resistance, fire resistance, etc.) or estimated/calculated (when the building predates the specifications)",

- specify the environmental profile defined by indicators enumerated by the norm XP P 01 020-3. This norm presents four categories of indicators. Firstly, the category "resources" focuses on the consumption of non-renewable energy resources, exhaustion of resources, consumption of water (both drinking and non-drinking water) and consumption of land area. Secondly, the category "air" relates to air pollution, climate change, atmospheric acidification, photochemical ozone formation and the destruction of the stratospheric ozone layer. Thirdly, the category "water" lists water pollution and eutrophication. Fourth and finally, the category "waste" is subdivided into hazardous, non-hazardous, inert and radioactive waste,

- enumerate the safety and comfort characteristics.

EXAMPLE.–"L'immobilier certifié durable – HQE Construction: le passage obligé" (Certified sustainable construction – HQE construction: the obligatory switch), *Transports & Logistique*, March 2012, p. 41-42.

---

11 http://assohqe.org/hqe/IMG/pdf/Referentiels2010.pdf.

"In France, Certivéa, the association created in 2006 by the *Centre Scientifique et Technique du Bâtiment* (*CSTB* – Scientific and Technical Center for Construction), has, in collaboration with Afilog, delivered forty "HQE Construction" certifications to France's contracting authorities and promoters for their creations of logistical depots. [...] The regulation *NF Bâtiments Tertiaires – Démarche HQE® "plateforme logistique"* is built, in particular, around three aspects. Location – and more generally, fitting in with the immediate surroundings – is the first of these. What do we do? We favor multimodal operations and monitor the flows of consumption and pollution. The second aspect relates to the improvement of working conditions, e.g. by using natural light. The third relates to the reduction of energy consumption, with a targeted reduction of 15% per m^2 from the carbon footprint."

2.5.4. *Evaluations of greenhouse gas (GHG) emissions*

Amongst the directives, approaches, etc., related to GrSCM, evaluations of the volume of emissions of greenhouse gases (GHGs) are of huge importance. Such evaluations take place both at the level of the industrial process as the use of habitat or indeed of transport.

2.5.4.1. *The Bilan Carbone® (carbon balance/footprint) method*

In France, the Bilan Carbone® method (carbon balance) is a normative framework. This approach was initiated by ADEME in 2003. It represents "a method for calculating greenhouse gas emissions reported in equivalent carbon or equivalent CO_2 of an industrial or administrative activity".[12] Its aim is to evaluate greenhouse gas emissions generated either directly or indirectly by the activity in question. Thus, here we consider emissions both and off the site under investigation, and before and after the occurrence of the activity being audited. The elements which are taken into account are "direct uses of energy, emissions due to procedures (leaks...), transport (goods, supplies, employees, visitors...), emissions relating

12 www.ademe.fr/bretagne/telechargement/CDC-bilan-des-emissions-de-gaz-a-effet-de-serre-d-une-entreprise-industrielle-ou-tertiaire.pdf

to incoming materials, to the management of waste products, to the end-of-life processing of the packaging, to the treatment of used water and the emissions from the manufacture of immobilized goods (buildings, vehicles, computer hardware) and to tertiary services other than transport."

EXAMPLE.–"McDonald's se taille une logistique durable sur mesure" (McDonald's is creating a made-to-order sustainable supply chain management policy), wk-transport-logistique.fr, April 2012, p. 40-41.[13]

> This partnership enables McDonald's, in particular, to coordinate as best possible the 215 suppliers with the 1100 references who manage all of the needs of the restaurants (food, packaging, consumables) by reducing distances as much as possible to obtain intervention radii of 150 kilometers on average.
>
> Thanks to this exclusive partnership, the chain has been able to drastically reduce the carbon balance of its logistical activities: between 2005 and 2010, McDonald's share of the total carbon balance shrank from 8% to 6%, and the restaurant's greenhouse gas emissions fell by 34.7% over the same period.

The Bilan Carbone® method can only be used by someone who has undergone the training offered by ADEME. It is a so-called "monocriterion" method, as it is based only on the criterion of "greenhouse gas emissions", taking account of the following gases:

– CO_2: carbon dioxide, which results from the combustion of fossil fuels – e.g. petroleum, coal and gas;

– CH_4: methane, which results from organic decomposition from anaerobic environment (an environment with oxygen), pyrolysis (decomposition of an organic compound by heat to obtain other products), etc.;

---

13 www.stamex.fr/mc-donald-s-met-sa-logistique-au-vert_129_49.html.

– N_2O: nitrogen protoxide, whose sources include nitrogenous fertilizers or the chemical industry;

– O_3: ozone, which corresponds to the photo-reaction of CH_4 (methane) and NO_x (nitrogen oxide);

– CFCs: chlorofluorocarbons, associated with coolant gases and industrial processes;

–SF_6: sulfur hexafluoride, produced by metallurgy and electronic compounds.

2.5.4.2. *Other methods*

In 2010, Depoers put forward a study on the consolidation of environmental data, and particularly of the volume of greenhouse gas emissions. For this purpose, the author considered three approaches:

– accounting, with the IAS-IFRS standards;

– Global Reporting Initiative (GRI);

– operational control.

The International Accounting Standards (IAS) or International Financial Reporting Standards (IFRS) have been in place since 1973, initially under the name of IAS, and then changing to IFRS in 2001. Generally combined under the appellation IAS-IFRS, these standards constitute referential frameworks for all listed companies. The criteria for consolidation of environmental data are based on the "entities integrated globally and proportionally" (p. 136).

The Global Reporting Initiative is an optional framework for sustainable development. It emerged in the United States in 1997 at the initiative of the "United Nations Environment Program" (UNEP). The GRI fits in with the environmental reports that companies publish or have to publish.[14] As regards the consolidation of environmental

---

14 It should be noted that since the introduction of the law on *nouvelles régulations économiques* (NRE – New Economic Regulations) in 2001, listed companies are obliged to include the environmental and social data relating to their activity in their economic reports.

data, it is "the concepts of control and notable influence and/or concept of significant impact on the environment" (p. 136) which are considered.

The approach of operational control is part of the Greenhouse Gas Protocol.[15] Promulgated in 1998, this protocol was jointly instituted by the World Business Council for Sustainable Development (WBCSD) and the World Resources Institute (WRI). Operational control over a type of source of GHG emissions is to be understood when an entity "(1) holds the majority of voting rights on the administrative board of the company, (2) has the authorization to operate delivered by the administration, and (3) by virtue of the terms and conditions of the contract regulating the right to operate the type of source concerned" (p. 135).

Depending on the type of approaches chosen, account will be taken of direct and indirect GHG emissions. So-called direct emissions relate to "emissions stemming from processes or machinery owned or controlled by the company" and so-called indirect emissions relate to "emissions connected with the company's activity, but from sites or operations owned or controlled by an entity other than that company" (p. 136). Thus, the three approaches (IAS-IFRS, GRI and operational control) all take account of direct emissions. However, as regards indirect emissions, IAS-IFRS does not take account of this type of emissions, GRI takes them into account if there is a significant impact and the operational approach takes them into account, considering that it is "wise to indicate the amounts of electricity and heat energy purchased and consumed" (p. 136).

2.6. Conclusion

A great deal of ink and page-space has been devoted, in recent years, to the topic of "green". The rarefaction of fossil materials, the environmental damage relating to pollution and to non-treated waste

---

15 Depoers [DEP 10] also cites the *Equity Share* approach as a second evaluatory approach.

are the subject of much debate. As it is, it will continue to become increasingly necessary, or even obligatory, to devise processes of production which reflect "eco" practices in the sense of ecology. For instance, before long, the holding of ISO 14001 status will soon become imperative in order to respond to the calls for bids made by clients. Similarly, limiting greenhouse gas emissions will involve the design of increasingly "clean" dwellings, and less pollutant means of transport… All these elements have a bearing on a sustainable supply chain.

2.7. Appendix

Respect de l'environnement: l'inventaire des bonnes pratiques (Respect for the environment: an inventory of good practices)
Les Echos, Tuesday 24 March 2009, p. 30
Dossier SITL temps réel (Salon international des solutions logistiques)
(Real-time dossier from the SITL (international logistical solutions trade show)). Author: O.N.

This (non-exhaustive) inventory, drawn up on the basis of interviews and numerous studies (AFT-Iftim, Ilec, etc.) has been validated by ADEME and *Club DEMETER Environnement et Logistique*, which brings together supply chain managers from manufacturers, distributors and managers. The President of Club DEMETER, Jean-Michel Rothier, expressed his appreciation (with, granted, a certain risk of subjectivity) and made positive comments about each of the projects (the number 1 represents the greatest environmental and/or economic benefit).It results from this that the organization, methods and processes are more important than the technologies.

Categories	Benefit
1. In advance of the transport phase (upstream), design products differently	
– favor eco-design of products and packaging and shelf-ready packaging (SRP)	3
– eliminate or reduce packaging and unnecessary over-packaging, seek out reusable packaging, trays or palettes	2
– lighten and compact products which can be lightened and compacted	3
– improve the lifetime of products	4
– improve the quality of products so as to limit after-sales flows	4
2. Reorganize the company's production and purchasing	
– take account of all the cost variables of long-distance supply and production (prolonged transport times, increased stock inventories, but also delays, less predictability, more difficult monitoring and poorer quality, increased spending on business travel, more frequent use of transport or of breakdown repairers, etc.)	2
– in the medium term, anticipate high petrol prices and quotas and taxes on CO_2so as to be able to imagine drastically different scenarios	1
– relocate distant production sites to closer sites →this decision will require extensive investment	2
– de-specialize factories serving large geographical areas in favor of more localized production near to the consumption basins →however, this will lead to an increased industrial cost	2
– rethink commercial provisions which increase the number of references and promotional hiccoughs by complicating and de-organizing the transport and logistical chains→i.e. bring the supply chain back to the heart of the process	2
– group your purchases with those suppliers who are closest to you	4
– seek out the closest competitive suppliers	?
– give thought to keeping the circuits between the producer and the consumer as short as possible	2
– with large-scale importing, group together and consolidate flows upon leaving the zones of exportation	3

Table 2.5. *Inventory of good practices in terms of the environment (source: Les Echos, 2009)*

3. Organize your logistics and transport better	
– put in place a structure for competent supply chain management with transversal power (with regard to purchasing, production and advertising), train staff at all levels in the company → in the process!	1
– audit your practices, call on specialist advice	4
– plan your production and storing as accurately as possible (too much or not enough stock causes costs and non-optimal movements)	1
– lend flexibility to your logistical organization in order to be able to react to events (market variations, change in the company's remit, change in suppliers, etc.)	1
– equip yourself with computer tools that are apt, in particular, for making predictions, planning and integrated management, and for creating a dashboard representation of your logistical costs and emissions	2
– weed out wastage in all links in the chain (better known as lean logistics)	2
– group together operations on the same site rather than transporting tools from one subcontractor to another (e.g. printing + binding + distribution in book publishing)	2
– assess your itineraries as closely as possible (reduce the miles covered, avoid backlogs, equip your fleets with tracking devices)	1
– organize your reverse logistics (packaging, old products, repairs, exchanges, unsold stock, etc.)	2
– plan your transport purchases in the long term	3
– find a balance between too-frequent playing off of transporters and logistics providers against one another and long-term partnerships	4
– avoid the excesses of surplus tension in the chain and of "zero stock" so as to avoid catastrophic stock breakdowns (lost sales, costly last-minute orders and transport, etc.)	2
– prepare ISO 14000certification (environmental management)	4
4. Combat unnecessary miles and speeds	
– rethink the whole of your supply chain so as to reduce the tons per mile	1

Table 2.5. (continued) *Inventory of good practices in terms of the environment (source: Les Echos, 2009)*

– improve the planning and itineraries of the delivery rounds made by the trucks	1
– weed out miles covered by empty vehicles	1
– segment your supply intakes and dispatches according to their actual degrees of urgency	1
– avoid express transport or last-minute air freight, consequences of poor prediction and poor organization of the supply chain	1
– design your supply chain so as to use those modes of transport which are slower but more consolidated, more economic and less heavy emitters of CO_2 (rail, river, sea) →the opportunities of multimodal transport	2
– condense and consolidate transport to reduce the number of vehicles in circulation	1
– adapt your general sales conditions so as to avoid the increased cost of distribution which is too fast in relation to the market requirements	1
– restrict the power of the engines of the vehicles so as to cap their speed	3
5. Reduce consumption of the means of transport	
– favor newer vehicles, which are green and clean, which consume less or use renewable energies	2
– improve the vehicles in technical terms (restriction of engines, aerodynamic accessories, tires, automatic gearboxes, self-cooling engines, etc.)	2
– adjust your fleet of vehicles to your genuine needs, or subcontract the transport to a rental agency	4
– optimize the upkeep and management of vehicle fleets	4
– train drivers in eco-driving and in behavior (switching off engines when stopped, use of air conditioning, etc.)	3
– support research and innovation in terms of clean vehicles	2
– rewrite the technical specifications and calls for bids to include environmental criteria	2
– limit unnecessary movement of personnel (teleconferencing, tele-training, cooperative tools, carpooling, etc.)	2

Table 2.5. (continued) *Inventory of good practices in terms of the environment (source: Les Echos, 2009)*

– revise your mailing lists so as to reduce the dispatching of mail and mistaken returns	4
– dematerialize invoices and carry out payments online to reduce expenses on paper, flows of mail and waste management	3
6. Consolidate transport operations so as to reduce unitary costs and emissions	
– increase the density of filling (more compact packaging, higher palettes and better filled trucks, combination of heavy and light goods)	1
– increase the capacity of the transport units (e.g. layers one on top of another in a truck, or higher palettes)	1
– make the transition to larger vehicles (44 tons or 60 tons if they are authorized and there is no alternative mode)	1
– group together your purchases, your suppliers, your deliveries, etc.	3
– increase the size of your orders and your lots	3
– combat the phenomenon of the trucks returning empty, by organizing return freight	1
– consider a switch in mode of transport: favor modes of mass transport (e.g. rail or river), by arranging with other carriers or with your logistics providers, or indeed by adjusting the rhythms of your orders and dispatches	2
7. Mutualize and cooperate between carriers and providers	
– identify and establish relationships with industrialists, distributors and logistics providers with whom you can work to concentrate flows using shared means (transport, platforms)	1
– put in place procedures and computer tools for collaboration, information-sharing, Efficient Consumer Response (ECR), electronic data interchange (EDI), etc.	1
– organize pooling (filling of trucks by multiple orders), multidrop (combination of small deliveries to nearby customers), multipick (concentration of deliveries from multiple suppliers), etc.	1
– use shared logistical platforms	1

Table 2.5. (continued) *Inventory of good practices in terms of the environment (source: Les Echos, 2009)*

8. Optimize your storage and distribution platforms	
– without relying on modes, adapt the organization of the warehouses to the specific needs of your supply chain; avoid a proliferation of hubs, platforms, shops and depots which increase the dispersity of cargos and detours on delivery rounds; or, conversely, too centralized platforms (national or continental) which increase delays and delivery times	2
– lend favor to new high environmental quality (HQE) platforms, which cost more to rent but which save energy	3
– rent rather than invest in a platform so as to preserve greater flexibility to adapt to your own network and your own markets	5
9. Improve the later links in your distribution chain	
– favor e-commerce, whose energy and environmental balance is more attractive than deliveries to superstores →true, but this is limited	3
– organize delivery rounds which avoid rush hour in cities (e.g. by delivering stock to stores very early in the morning)	1
– use "soft" modes of transport (electric vehicles, electrically-assisted bicycles, etc.) for small urban distances →too marginal	5
– avoid multiple deliveries to the same customer	3
– create relay points to limit the number of vehicles and the mileage covered	2
– use river ports and railway depots to consolidate incoming flows	1
10. Communicate in order to lend value to pioneering approaches	
– put in place dashboard representations and performance indicators	1
– set objectives for the reduction of CO_2 and subscribe to voluntary business commitments (charters)	1
– communicate internally (a highly mobilizing subject) and externally (a highly valorizing subject)	1

Table 2.5. (continued) *Inventory of good practices in terms of the environment (source: Les Echos, 2009)*

Chapter 3

The Social/Societal Aspect of Sustainable Supply Chain Management

3.1. Introduction

The performance of any and every activity is a crucial element in all managerial thinking. Thus, amongst the elements which need to be considered, we have to understand the behavior of individuals. Indeed, at some levels of the organizational strata, every individual has a key part to play. In fact, there are many questions which arise in organizational management. Which interlocutors should we be talking to? Is it appropriate to uniformly inform all the employees in an organization and each organization within the supply chain? Should we envisage transmitting information *en masse*, at the risk of "blurring" the strategic vision instead of accentuating its crucial points? It must be remembered that an excess of information risks obscuring information, as it is commonly said: thus, we might worry that the declarations transmitted will become confused and therefore unreliable because of their over-extensive distribution. Conversely, if we choose to "categorize the information", then those people who belong to the same organizational unit but do not have access to the same information may feel the sentiment of iniquity and lack of recognition for a shared project [MOR 04b].

However, can we not speak of the performance of organizations without looking at social and societal norms? In that sense, it is crucial to reflect on the link between human resource management and logistics. Therefore, the human dimensions are often highlighted in supply chain management and green supply chain management. For instance, in SCM, certain authors speak of the general behavioral context which needs to be (re)conciled with behavioral and managerial components, a common culture, definition of roles, risk- and reward-sharing, the work patterns, conflict reduction, training or ideas boxes [GRL 95; COO 97; BOW 99; LAM 98; LAM 05; MEN 01; MIN 04]. In GrSCM, Srivastava [SRI 07] stresses occupational health and safety. Also in 2007, in a layman's-terms article about the link between SCM and human resource management, Emeric Levy of CPIM-APICS ([LEV 07] in *Logistiques Magazine*, n° 222) notes that if the human factor is ignored, organizations become stuck in their process of evolution because of inappropriate organization, failing processes, unidentified dysfunctions and demobilized teams.

Therefore, a publication on the advantage of ensuring "social/societal supply chain management", in the same vein as green supply chain management which takes account of environmental thinking in SCM, is opportune [MOR 08]. Indeed, it is necessary to define a social/societal SCM which is able to deal with the ever-increasing need for talent, but also to combat a certain form of workplace blackmail, which involves "brandishing a sword of Damocles" by the practice of delocalization to countries with low wage expenditures.

3.1.1. *In favor of the definition of social/societal supply chain management*

A concept which emerged in the 1950s to limit certain unethical trajectories of the organizational economic system, "corporate social responsibility" (CSR) specialized in the social aspect by promoting voluntary integration of social and ecological concerns into industrial and commercial activities [IGA 02]. In the same vein, we speak of socially responsible investing (SRI)[1] which, besides the criteria relating to financial profit, integrates ethical, governmental, social and environmental aspects into any investment.

> Definition of corporate social responsibility (taken from [IGA 02]:
>
> "Most definitions for corporate social responsibility describe it as the voluntary integration of social and ecological concerns into industrial and commercial activities.
>
> For that reason, we sometimes speak of the concept of the threefold result: a good performance then becomes economic, social and environmental. We can also consider the company as being anchored in society, i.e. in constant interaction with social groups who have differing expectations."
>
> Introduction, p. 13
>
> "Corporate social responsibility (CSR) is companies' voluntary integration of social and ecological concerns into their commercial activities and their relationships with all the internal and external stakeholders (shareholders, staff, customers, suppliers and partners, human collectives, etc.) in order to fully comply with the applicable legal obligations and invest in human capital and in the environment" (European Commission Green Paper – 18 July 2001).
>
> Introduction, p. 15

1 The *Association française de la Gestion financière* (French Association for Financial Management) and the *Forum pour l'Investissement responsable* (Responsible Investing Forum) gave a definition for SRI in their code of transparency: "Application of the principles of sustainable development to investing. An approach consisting of systematically taking account of the three dimensions, which are the environment, social/societal factors and governance (ESG) in their usual financial criteria. The means of implementation may take multiple forms based on positive selection, exclusion or both at once, with the whole process integrating dialog with the providers, if need be" (Translated from p.5 of the European Code, to be found at www.afg.asso.fr/index.php?option...en).

Similarly, governments have launched an appeal to promote sustainable modes of production and consumption, to improve goods and services whilst decreasing the impacts on the environment and on health. This thinking is partly to be found in the definition of eco-socio-design, which is aimed at providing goods and services that satisfy people's needs and that contribute to the quality of life, whilst progressively reducing environmental nuisance and the negative social impacts throughout the lifecycle of the product [VAN 08; MOR 08].

However, as we can see, these various works often combine social and environmental aspects. It therefore seems wise to devise – as far as possible – a form of social/societal supply chain management which focuses only on the social and societal aspects involved.

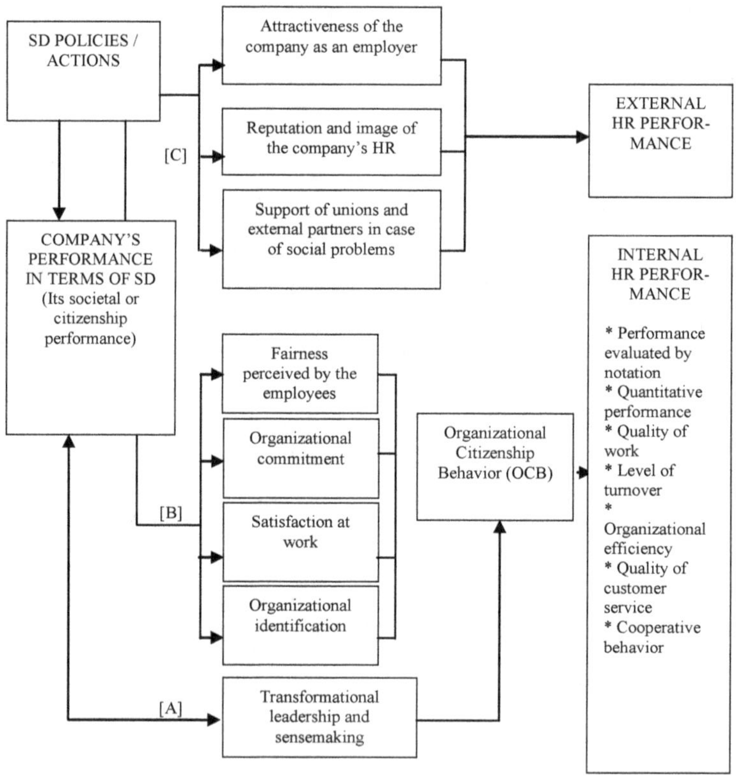

Figure 3.1. *The policy of sustainable development as a lever to develop the performance of human resources (source:[GON 06]): [A]: leadership, [B]: citizenship behavior within the organization; [C]: company's external reputation; SD = sustainable development*

Consequently, in addition to supply chain management and green supply chain management, it is helpful to outline a social/societal supply chain management, which aims to pursue a strategic management of the quality of life for human beings. As regards the determination of the constitutive elements of social/societal SCM, a parallel can be drawn by the work of Gond [GON 06] relating to the determination of appropriate criteria to improve the policy of sustainable development by developing the performance of human resources (see Figure 3.1).

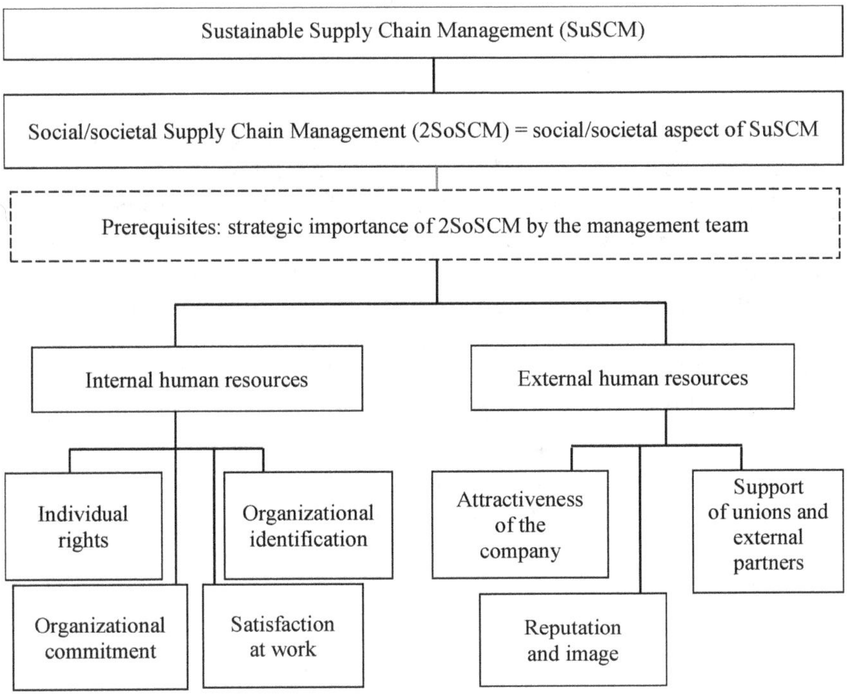

Figure 3.2. *The social/societal aspect of sustainable supply chain management*

In the context of this book, we modify this diagram somewhat, so as to "stick to" figures relating to economic and environmental dimensions of sustainable SCM. Thus, before any decision/action, we place "the strategic importance of social/societal SCM" and develop the internal and external human resources in the context of logistics (see Figure 3.2).

3.2. Internal human resources

The tools, principles, approaches, etc., relating to internal human resource management are many [MOU 12]. Here, we shall present and discuss a few of these, which are deemed to be crucial, and we shall highlight their connections with logistics as far as possible.

3.2.1. *Individual rights*

At the heart of the fairness perceived by the employees, we recognized the rights of individuals. On an international scale, the SA8000 standard (SA stands for Social Accountability) and its nine chapters on the respect of fundamental labor rights is recognized as being necessary regarding corporate social responsibility within supply chains [CIL 08]. However, in France, for instance, this private framework holds no real interest when we consider that the codes of conduct and the legal regulations largely cover the elements of this norm [DAM 04]. Thus, in a survey put to logisticians in France, Senkel and Koleva [SEN 08] noted that on this level, logisticians do not refer to SA8000, with a little more (18%) referring to the OHSAS standard, which we shall discuss later on.

3.2.1.1. *The SA8000 standard*

While SA8000 is not a necessity in companies in France, it is necessary to be aware of it in light of the different exchanges within a supply chain, including exchanges with foreign structures.

The SA8000 standard was drawn up in 1997 by an American organization: SAI (Social Accountability International). It relies upon human rights reference texts. It is based on the conventions signed by the International Labor Organization (ILO), the Universal Declaration of Human Rights published by the United Nations (UN) and on the UN Convention on the Rights of the Child.

The SA8000 standard is available to companies wishing to guarantee that their workers' fundamental rights will be respected. Its aim is to ensure that work is carried out in a safe and decent way. It is subdivided into nine areas: the first eight relate to the social aspect, and the ninth focuses on the management system needing to be implemented in order to ensure conformity with this standard.

1. Child labor: the SA8000 standard is based on ILO conventions n° 138 and 182. On this point, the standard stipulates that the company must not encourage work by children under 15 years of age. In addition, the company must take all possible measures to ensure that children have the opportunity to go to school, offering them – as far as possible – any means necessary to facilitate this schooling by way of a "remediation[2] procedure".

2. Forced and compulsory labor: SA8000 stipulates that the company shall not engage in or support the use of forced or compulsory labor. This is characterized by its neither accepting nor requiring personnel to lodge identification papers upon commencing employment. Consequently, each worker must be engaged on a voluntary basis, must have an employment contract and benefit from the protection of their employer, in conformance with the regulation in force locally.

3. Health and safety: the company must ensure that all workers have a safe and healthy working environment. This means that it must take adequate preventive measures to avoid any danger and/or accident which could damage their health.

4. Freedom of association and the right to collective bargaining: the company must respect the right of all workers to form or join a trade union of their choice and the right to collective bargaining. This right is fundamental for workers, and the company must do everything within its power to exercise and/or facilitate the application of this right.

2 Remediation consists of putting in place all possible means to resolve difficulties with schooling/learning (e.g. personal tutoring, resits, etc.).

5. Discrimination: the company must preclude all forms of discrimination. This must be understood in the broadest sense of the term. Thus, employees must not be subject to discrimination on the basis of their race, caste, national or social origin, religion, disability, gender, sexual orientation, union membership or political affiliation; nor remuneration, compensation, access to training, promotion, firing and retirement, etc. With this in mind, any act, behavior, gesture, words and/or physical contact which implies an obligation, threat, abuse or form of exploitation may be qualified as "discrimination". This domain also covers equal remuneration.

6. Disciplinary practices: the company must prevent all punishment, be it corporal or verbal. Moral harassment must be proscribed, as must any pecuniary sanction.

7. Working hours: SA8000 refers to the legal depositions in force in the countries in question.

8. Remuneration: workers should be guaranteed a salary which is able, at the very least, to cover all their basic needs (food, lodgings, clothing and healthcare) without having to work overtime to receive it.

9. Management systems: this point is extremely important because it stipulates that all companies must have documentary proof of its management system and of the procedures implemented to respond to the requirements of the SA8000 standard. The goal is that in case of an audit, this documentation should reflect true and continuous practices.

3.2.1.2. *The codes of conduct*

It was in the late 1980s that the codes of conduct emerged in France. They became widely known during the 1990s. The codes of conduct are intended to highlight the values of a company. According to Mercier [MER 02], "the codes of conduct contain primarily a deontological dimension (it would be more appropriate to speak of professional ethics, or of internal deontology): they provide a prescriptive formulation of the rules and duties governing the company's activity. These codes are binding within the company, and are imposed on the employees when facing a certain number of situations". By extension, these codes may affect certain stakeholders in the supply chain, such as suppliers, in guaranteeing the company's good image with employees and/or consumers. Although they have no real legal foundation, the codes of conduct are classified as so-called "soft law" [ETR 07].

3.2.1.3. *Social balance*

In France, a decree of application of the law of 12 July 1977, dating from 8 December 1977, stipulates that companies with 300 or more employees must put in place a social balance. This is an analytical tool which uses numerical indicators to evaluate the social data available within the company. It includes four categories which should help take stock of the company's situation (article L. 438-4 of the Labor Code, version in force on 20 August 2010).[3] We cite:

– employment, including in-house personnel, external workers, resignations, promotions, unemployment, disabled staff, absenteeism;

– remuneration and additional charges, including the amount of remunerations, the hierarchy of remunerations, the way in which remunerations are calculated, additional costs, overall cost of salary, salarial contribution of the employees;

– health and safety conditions, listing workplace and transport accidents, the distribution of accidents from material elements, professional illnesses, the health and safety committee, expenditure on safety;

– other working conditions, listing the duration and planning of work, the organization and content of the work, the physical working conditions, the transformation of the working organization, work place medicine, unfit workers.

3 http://droit-finances.commentcamarche.net/legifrance/61-code-du-travail/164777/liste-des-informations-figurant-dans-le-bilan-social-prevues-a-l-article-l-438-4

3.2.2. *Organizational commitment*

Organizational commitment can be appreciated in a transactional exchange. A number of dimensions appear in a transactional exchange, which brings together the employee and the employer [DEM 97]. This relation focuses on the concepts of competences and capabilities, motivation, training provided by education and/or by the company, and on evaluating consensual investment (the remuneration-to-involvement ratio). Below, we explain a few of these points.

3.2.2.1. *The concept of competences and capabilities*

Competences can be appreciated on a variety of levels. Thus, we can list:

– individual competences (*in vivo*) which comprise knowledge and know-how, typical conduct, standard procedures, types of reasoning [DEM 94]. Le Boterf [LEB 95] subdivides the notion of knowledge and know-how. For this author, there are three types of knowledge and two types of know-how. Knowledge may be theoretical (knowing how to understand and interpret), procedural (knowing how to proceed) or experiential. As regards know-how, the types are: social (knowing how to behave, to conduct oneself) and cognitive (knowing how to process information, how to verbalize what one is doing, and knowing how to learn);

– relational competences (*in vitro*) which involve dealing with the existent and latent dangers of one's environment. For this, the company has to implement central competences throughout its entire range of products or services and in the various existing or new market sectors [HAM 95]. In 1990, Prahalad and Hamel [PRA 90] defined central competences as "core competencies are the collective learning in organizations, and involve how to coordinate diverse production skills and integrate multiple streams of technologies" (p. 82);

– professional competences are understood as "a unique combination of know-how and experience, which takes a long time to construct and is difficult to imitate" [MIL 91](p. 59), including basic technologies, industrial expertise, understanding of customers' behavior and needs, management of an industrial brand and knowledge of the distribution channels.

The case of GPEC (*Gestion Prévisionnelle des Emplois et des Compétences* – Projected Management of Jobs and Competences)

Projected management of jobs and competences (or *GPEC*) dates from the 1970s. It is to be understood as the appreciation of the needs and human resources held by a company. It is established on the basis of clearly-identified objectives, and requires involvement of the employee in his/her own project of professional development.

The French government put in place a three-yearly obligation as regards *GPEC*, as part of Law n° 2005-32 – Planning Act for Social Cohesion, 18 January 2005, version consolidated on 1 September 2010 – see www.legifrance.gouv.fr/affichTexte.do?cidTexte=JORFTEXT000000806166.

Capabilities are to be understood as the firm's ability to generate a competitive advantage for itself. The intended goal is to help employees understand and meet the clients' requirements by developing organizational capabilities which are "idiosyncratic and inimitable" [ULR 91]. For Milan [MIL 91], the term covers the capacity to reduce one's response time, acuity which consists of having a clear view of the markets and of one's competitors, the implementation of quality from a productive and behavioral point of view, multi-valence by team management and the development of intercultural exchanges, and the capacity to learn, which multiplies the aforementioned capabilities. In concrete terms, Stalk *et al.* [STA 92] hold that design of a strategy must be based on a "capability-based competition", which combines:

– rapidity, i.e. the capacity to react quickly to the demands of the customer and of the market, and incorporate new ideas and technologies into the products as swiftly as possible;

– coherence, understood as the ability to design a product which unfailingly satisfies the consumers' desires;

– the acuity to view the competitive environment clearly and anticipate and cater for the consumers' requirements;

– agility as the aptitude to simultaneously combine several different environments;

– innovation to generate new ideas and combine the existing elements to create new sources of value.

We cannot speak of a fundamental difference, as such, between the concepts of competence and capability, as capability is perceived more generally.

Le concept de compétence (The Concept of Competence), Chapter 2 (extracts and summaries), C. Lemoine [LEM 02]

"The concept of *competence* has dethroned the previous concepts of *aptitude*, *capability* and *qualification*, integrating the idea of mobilization, motivation, commitment and involvement, and taking up the baton of the old behaviorist reference of externally-observed behavior, or behavioral response."

Capability: "a set of recognized performances which may be defined by one or more referential frameworks (AFNOR standards, [AFN 94])";

Aptitude: "denotes the constitutional substrate of a capability, pre-existing it, which will depend on the natural development of the aptitude, educational training and exercise [PIE 73]";

The term *ability* encapsulates the notions of capability and aptitude.

Qualification: "official recognition, attested by a certificate, of aptitudes or level of training, expertise or theoretical/technical knowledge, acquired either in an educational institution or through professional experience".

In this arrangement, we see a clear convergence of terms with the practice of logistics and in the area of transport. Camman and Livolsi [CAM 08] put forward a map of capabilities/competences in the transport of goods. Indeed, "the increasing internationalization of operators and the development of the offer of door-to-door transport or of global logistics" mean that professionals in the transportation of goods (transport, transport organization, transport engineering[5]) have to improve both their haulage operations and the organization and management. This development of competences is defined on an internal and external level (see Table 3.1).

<table>
<tr><th colspan="2">ORGANIZATIONAL CAPABILITIES
Capability for integration of different professionals in the supply chain
Capacity for innovation</th></tr>
<tr><td>Internal development</td><td>Brand image/reputation
Action on the structure of the organization and modes of management (transversality): project teams, large account managers
Development of transversal management systems: control of management and information systems
Training: initiating a thought process of exploration at all levels
Communication: promulgating the strategic vision</td></tr>
<tr><th colspan="2">INTERFUNCTIONAL CAPABILITIES
Management of logistical chains and transport
Relationship skills – Customer service – Capacity for innovation</th></tr>
<tr><td>Internal development</td><td>Development of project teams to design innovative services (strategic and organizational surveillance) and management of customers' logistical organizations
Quality approaches (certification): improvement of performance
think tanks: developing the company's services, favoring collective learning, capitalizing and transferring knowledge between agencies, etc.</td></tr>
</table>

Table 3.1. *The competences involved in transport and modes of development (source: [CAM 08])*

5 Transport includes consolidation/deconsolidation operations, organization of delivery rounds and rental. Transport organization includes the commissioning/organization of transport, transit and customs.

FUNCTIONAL CAPABILITIES Human resource management – Marketing – Information system	
Internal development	*Development of marketing approaches*: particularly communication (enhancing value of the company's offering, image, reputation) *Development of HR policies* (training, recruitment, career management, evaluation): developing individual competences, involvement in exploitation and exploration, motivation, communication of the strategic vision *Information systems*
External development	*Information system*: alliances with software development companies: design and integration of optimization tools
ACTIVITIES AND CAPABILITIES SPECIFIC TO PROFESSIONAL FUNCTIONS (TRANSPORT OPERATIONS)	
Internal development	*Increase of transportation capacities* *Development of transport purchasing cells*: reduction of the panel of suppliers, reliability of contracts, reduction of cost of purchase, stabilization of certain transport schemas, control of externalized activities, etc. *Development of new modes of transport* whether or not associated with the primary transport: cost reduction, consideration of sustainable development issues *Activity-based information systems*: TMS (Transport Management Systems), traceability of lots, etc. *Development of networks of agencies on an international scale*
External development	*Transporter acquisition*: swelling the size of the fleet, integration / reintegration of physical operations into transport *Development on an international scale* by acquisition, acquisition of equity, etc. *Transport purchasing* Outsourcing of activities relating to the running of the transport network (purchasing and flow optimization) Acquisition of activities linked to the organization and control of transport Alliances, whether or not bolstered by partner companies whose activity lies in the organization and running of transport

Table 3.1. (continued) *The competences involved in transport and modes of development (source: [CAM 08])*

In 2009, Livolsi presented the profile of the supply chain manager (*Logistiques Magazine*, n° 241).

EXAMPLE.– *Supply Chain Manager: maîtriser les rouages de l'entreprise* (supply chain manager: controlling the cogwheels of the company), An Inquiry by L. Livolsi, *Logistiques Magazine*, n° 241, July-August, 2009, p. 54-58.

"The competences of a supply chain manager are characterized by:

– "more strategic activities of flow management, sales projections, forward planning of supply and realization of the industrial and commercial plan";

– "a supply chain manager, on average, manages a twenty-strong team. He must be capable of managing and working in an ERP environment, a tool which requires him to be able to understand the representation of the way in which the company is organized";

– generally around forty years of age: "companies minimize the risk of recruitment as far as possible by looking for professional maturity and already-gained experience in SCM";

– "a good profile is also characterized by good intercultural competences both at the international level and at the level of the different professional groups within the company. The candidate should also demonstrate organizational capacities, proving himself able to design a new organizational structure and lead the change. He must be able to manage and unite diverse teams, and to make apt use of managerial tools (dashboards, lean production, management control, etc.)";

– "It is a decision-making position, whose occupant has to demonstrate that he is contributing to the company's performance by generating added value".

3.2.2.2. *The concept of motivation and loyalty*

According to Louart [LOU 97], motivations express the "dynamic aspects of human behavior" and must be analyzed as an analytical framework rather than as a means of controlling behavior", i.e. that we can understand the meanings and personal representations, but that it is up to each individual to share them.

The quest for motivation and loyalty from one's employees is an element which must not be overlooked in the field of logistics. We can understand it by way of the notion of corporate social responsibility (CSR). Thus, Ciliberti *et al.* [CIL 08] speak of corporate social responsibility which, in a supply chain, involves two factors:

– conformity with the requirements by use of written documents and evaluation of the performance;

– reinforcement of the capacities, which is envisaged by way of awareness-raising and training.

Nevertheless, in an international supply chain, there are many problems that arise in terms of ensuring the legitimacy of CSR. Of these problems, we can cite cultural differences, language, poor customer interest, which "impose" a reasoning based on as low a price as possible, which is often inversely proportional to social legitimacy, the site of production (developing countries, corruption in certain countries), religion and problems with ICT because some people do not have personal computers – a crucial tool in trade.

3.2.3. *Organizational identification*

Organizational identification occurs "when an individual's beliefs about his organization become self-referent or self-defining" [PRA 98; HER 06]. In the eyes of Chedotel [CHE 04], organizational identification gives the employee the feeling of "belonging to the organization", and enables him to answer the question: "Who are we?" – a question which relates to the organizational identity. Thus, for instance, not identifying with the organization to which one belongs can – in extreme cases – drive one to suicide. Garcia [GAR 09] represents organizational identification by way of Figure 3.3.

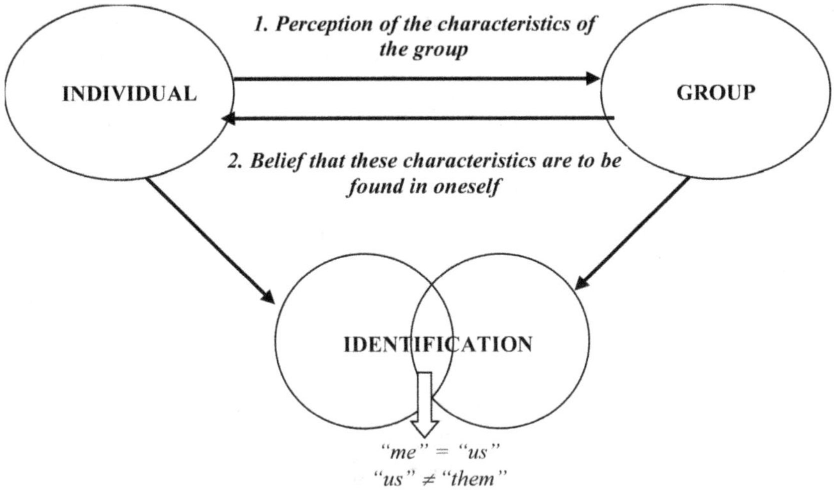

Figure 3.3. *Process of organizational identification (source: [GAR 09])*

3.2.4. *Job satisfaction*

Under the umbrella of "job satisfaction", we find autonomy, interest and working conditions, as well as career prospects [PAL 00]. Many of the examples currently being raised in professional articles in logistics are oriented in this direction: from the protection of employees in outsourcing contracts (*Logistiques Magazine*, October 2009), to valorization of work in warehouses: implementation of "voice picking", RFID and the "goods to man" principle (*Logistiques Magazine*, September 2009).

EXAMPLE. – Protection of employees: personnel transfer included in the outsourcing contract – "Les critères de réussite d'une externalisation", (The criteria for successful outsourcing), *Logistiques Magazine*, October 2009, n°243, p. 20-24.

In logistics, there is no standard legal contract. Thus, we can see the importance of making clear calls for bids which specify the different clauses in the contract, particularly those relating to the duration (three, six or nine years, of determinate or indeterminate duration), price (to the volume of activity), insurance (in case of a slump in activity) and exit clause from the contract. This "exit clause" is of fundamental importance, in that it must stipulate transfers of stock, information systems, materials, current leasing and above all, personnel.

EXAMPLE. – Facilitation of employees' work: valorization of work in warehouses – *"L'homme au cœur de l'entrepôt"*, ("The human heart of the warehouse"), *Logistiques Magazine*, September 2009, n 242, p. 40-43.

Warehousing may represent between 1/3 and 2/3 of a company's logistical costs. However, following a technical approach, optimization must now take account of the human dimension. Indeed, the work of warehouse operators is repetitive, difficult and not hugely motivating. These points are to be seen in an increase in professional illnesses and absenteeism of sometimes over 10%. Add to this a high rate of staff turnover, which leads to organizational problems. To make these people's jobs easier, many practices have been put in place, including order preparation by voice picking, use of RFID and tag readers or indeed automation in line with the "goods to man" principle. Similarly, specialized software packages have been implemented so as to optimize the distribution of the workload, compare the performance of teams and operators, determine the optimal composition of teams, monitor work through the distribution channel and calculate rewards (bonuses or re-timetabling).

EXAMPLE. – Facilitation of employees' work: valorization of work in warehouses – *"Des matériels innovants"* ("Innovative materials"), *L'entreprise*, March 2010, n 287, p. 118.

R&D is leaning towards the construction of "forklifts which are safer, less energy-hungry and less damaging to employees' health". Thus, we are witnessing the emergence of pallet trucks, which are automatically raised to a height of 80 cm, or forklift cockpits which adapt to the morphology of the operators.

The facilitation of employees' work is, therefore, a crucial element in a just-in-time logistics approach. Indeed, when the objective is to tend toward having a zero inventory from the supply stage, the role of human resource management is to facilitate employees' work by any means necessary, so that these employees do not (for instance) go on strike, which would break the flow [CLA 04].

The implementation of health-and-safety normative frameworks such as OHSAS 18001 is also part of job satisfaction.

3.2.4.1. *The OHSAS 18001 and ILO-OSH 2001 standards*

The OHSAS 18001 framework (OHSAS stands for Occupational Health and Safety Assessment Series) is a private framework. In 2007, it was reviewed, and became known as BS OHSAS 18001: 2007. This change altered its status from being a guideline to a national British standard (BS). It was with the view to corresponding as closely as possible to the ISO 9001 and 14001 standards that this new version was drafted. BS OHSAS 18001: 2007 is founded upon the following principles:

– "identify and control your health and safety risks;

– reduce the potential for accidents;

– help achieve compliance with health and safety legislation;

– improve your organization's performance"[6];

EXAMPLE. – *"Chariots élévateurs. Deux nouveaux sites certifiés OHSAS 18001 pour Toyota Material Handling"* (Forklifts: two new sites given OHSAS 18001 certification for Toyota Material Handling), *Supply Chain Magazine*, Newsletter 1422, 29 May 2012.

> "In the wake of the factories in Ancenis and Bologna, two more sites (production and separate parts) owned by the TMHE/BT Products AB group, obtained OHSAS 18001 certification last month (this is an H&S management framework). The factory in Mjölby

6 http://www.bsigroup.com/Documents/bs-ohsas-18001/resources/BSI-BSOHSAS18001-Product-Guide-UK-EN.pdf

> (Sweden) constitutes one of the world's largest sites for production of warehousing vehicles, with 76,000m^2 devoted to manufacture, and 1,450 employees."

The OHSAS 18001 framework was followed by the international framework ILO-OSH 2001, promulgated by the International Labor Organization – "Guidelines on Occupational Health and Safety Management Systems". This framework comprises five general principles:[7]

– policy, which deals with occupational health and safety policy, and with workers' participation;

– organization, relating to employer responsibility and obligations, competences and training in terms of health and safety, documentation of the occupational health and safety management system, communication of orders and procedures;

– planning and implementation of the system, including the definition of objectives and risk prevention;

– evaluation by monitoring and measurement of effectiveness, inquiries into the causes and effects of problems encountered, auditing, examination by the management team;

– action with a view to improvement, which involves preventive and corrective action, and continuous improvement.

3.3. External human resources

Remember, human resources relate to: the attractiveness of the company, its reputation and image, and the support of unions and external partners in case of problems.

7 www.ilo.org/wcmsp5/groups/public/---ed_protect/---protrav/---safework/documents/normativeinstrument/wcms_112581.pdf: Guidelines on Occupational Health and Safety Management Systems.

3.3.1. *Attractiveness of the company*

The attractiveness of the company relates (may relate) to the learning opportunities which an organization offers. Thus, the notions of organizational learning are increasingly echoed by practitioners and researchers, and many aspects of it have been discussed in print.

Thus, according to Probst and Büchel [PRO 95], organizational learning is to be envisaged as "the enlargement and the change of the values system and knowledge system, improvement of problem-solving capabilities and capacity for action and alteration of the reference framework shared between all individuals within an organization" (p. 16). Although this notion of organizational learning is sometimes viewed as being similar to the notion of a learning organization [SEN 90], the latter seems to be more globalizing, integrating group synergy within the organization. Organizational learning, which relates more to the individual domain, therefore constitutes one of the facets of the learning organization.

In itself, it is the capacity to learn from their own experience which enables companies to obtain better products and greater profits. It is through collective and cumulative learning, and an ability to constantly renew its approaches, that an organization increases its ability to take advantage of all the changes occurring in its environment (Lampel in Mintzberg *et al.*, [MIN 99]).

3.3.2. *Reputation and image*

The strategic importance of social/societal SCM involves the fundamental role played by a company's management team (MT), but also by the MT of every department within an organization. According to Boiral [BOR 04], a good corporate image is influenced by the implementation of standards, such as the environmental standard ISO 14001 – or at least, such is the opinion of the Canadian companies surveyed by that author.

EXAMPLE. – Improvement of image by use of ISO 14001 – *Mettre en œuvre l'ISO 14001: de la quête de légitimité à l'émergence d'un "mythe rationnel"* (Implementing ISO 14001: from the quest for legitimacy to the emergence of a 'rational myth'"), [BOI 04].

> ISO 14001 certification is of crucial importance for a factory's "green" image. Given that the factory is located within city limits, it is important to project an image which proves that we are actively managing the environment (an employee, case-study 2).

> We have a part to play in society. We want to be an example of a good corporate citizen, because we are a focal point at regional level. I believe the paste and paper industry has something to prove in terms of the environment, because public opinion is not overly positive at that level. The adoption of a standard such as ISO 14001 shows that we are moving in the right direction (a manager, case-study 7).

Traceability also appears to be an important factor in the image of a company, or indeed of a country [KAR 10]. Thus, in her study of the factors in the adoption of a traceability approach in the date industry in Tunisia, Meriam Karâa points out that traceability – while it is, in the short term, complicated to implement on the financial and organizational level – fosters a good brand image for the future, particularly in the eyes of the end consumer.

EXAMPLE. – The role of traceability in corporate image – extract from M. Karâa's doctoral thesis: "Les déterminants de l'adoption de la traçabilité par les entreprises de conditionnement de dattes en Tunisie" (The determining factors in the adoption of traceability by date-processing companies in Tunisia), thesis in management sciences, Aix-Marseille II, 2010.

> In addition to the economic gain that traceability can offer, we felt it to be a way of improving our image. Between a company with traceability in place for its products and another which sells dates with no

traceability, which do you think would have the greater advantage and be favored by the customers? Of course, it would be the one with traceability in place (Head of Purchasing, Packaging and Other, Date processing company A).

3.3.3. *The support of unions and external partners*

External partners are of crucial importance in a supply chain, as we have already seen. We have mentioned the roles of the state or of Europe, in the promulgation of decrees, such as the decree on the processing of waste electrical and electronic equipment or on the increasingly advanced re-engineering offered by logistics providers. In the supply chain, we must forget neither the suppliers nor the final customers. In this section, we present unions, which affect the management of an organization as a whole. Here we should also mention the impact of socially responsible consumption, which promotes so-called “ethical” purchasing, and plays a role in the logistical approach (e.g. impact on transport). Finally, we can cite the ISO 26000 standard and the societal balance, of which, all the actors – including logisticians – need to be aware.

3.3.3.1. *Trade unions*

Although they are connected to the company, trade unions are “external actors” whose role is to assist employees and managers in the approaches and actions that are undertaken. Hence, they are actors whose influence is important, e.g. in conflict management.

3.3.3.2. *Socially responsible consumption*

Socially responsible consumption (SRC) aims to “take account of the public consequences of private consumption” [FRA 06]. In itself, SRC is part of environmental logistical thinking – if only, for instance, by the choice of favoring purchasing in the same country, which has an impact in terms of reducing CO_2 emissions.

EXAMPLE. – *Réduire notre impact logistique* (Reducing the impact of our logistics), Carrefour[8]

> "By promoting responsible consumption: by offering energy-saving products and raising awareness of the advantages to such products, developing ranges of eco-labeled products, encouraging the reuse of baskets and eco-bags and reducing the number of free disposable plastic bags."

François-Lecompte and Valette-Florence [FRA 06] propose a measuring scale comprising five factors in socially responsible consumption. This measuring scale reflects the behavior of the end consumer, and therefore – all other things being equal – helps companies in their day-to-day practices. These five factors are presented below, beginning with the factor deemed the most important by the respondents (sample of 1247 individuals):

– products-share purchasing, which includes four items of action/behavior – namely:

- "buying products from which part of the sale price goes towards a humanitarian cause,

- buying products from which the money goes to developing countries,

- buying products from which part of the sale price goes to a good cause,

- buying Fairtrade products (Fairtrade is a scheme which guarantees a decent level of living to small producers in developing countries)";

– organization, which includes five items of action/behavior – namely:

7 www.carrefour.com/cdc/commerce-responsable/notre-engagement-pour-l-environnement/reduire-notre-impact-logistique/logistique.html?com.carrefour.cdc.print.cart.force.empty=true, consulted November 2012.

- "not buying from companies or traders who have close ties to organizations such as the mafia or sects,

- not buying products made by companies that use child labor,

- not buying from companies whose practices toward their employees are not respectful,

- not buying from companies or traders who have close ties to political parties of which I do not approve,

- not buying from companies that pollute heavily";

– small trade, which includes four items of action/behavior – namely:

- "avoiding buying everything from large-scale supermarkets,

- buying from small traders (bakers, butchers, bookshops, etc.) as often as possible,

- keeping small businesses in my area alive by giving them my custom,

- going to a market to support small producers of fruit and vegetables";

– the origin of the product, which includes four items of action/behavior – namely:

- "when I have the choice between a local product and a product manufactured elsewhere in the world, choosing the local product,

- tending to buy locally-made cars,

- buying fruit and vegetables grown in my home country,

- buying products made in my own region";

– the volume of consumption, which includes three items of action/behavior – namely:

- "limiting my consumption to that which I really need,

- not over-consuming in general,

- not buying products that I can make for myself."

3.3.3.3. *The standards and tools associated with external partners*

3.3.3.3.1. ISO 26000

Initiated in 2001, the International Standard ISO 26000 on corporate societal responsibility was put in place to deal with the concerns of consumer organizations about the practices of multinationals which have consequences for the working and living conditions of the populace. It should be noted that five countries voted against the application of this text: Cuba, India, Luxembourg, Turkey and the United States.

This document is a guide, and therefore cannot be considered as a certification. ISO 26000 deals with seven main issues:

– the running of the organization,

– human rights,

– relationships and working conditions,

– the environment,

– loyalty of practices (corruption, competition, property, etc.),

– consumer relations,

– communities and local development.

3.3.3.3.2. Societal balance

Unlike the social balance, the societal balance is not obligatory. It has a place here because, although it refers to the employee (social balance), its approach is intended to be broader, including other stakeholders in the entrepreneurial project. The societal balance was developed in 1996 by the *Centre des jeunes dirigeants et des acteurs de l'économie sociale* (CJDES – Center for Young Managers and Actors in the Social Economy).

The societal balance comprises nine major domains and fifteen criteria for appreciation. The nine major domains are:

– products/services and services-customer relations;

– economic management;

– prospective-innovation-anticipation;

– organization of work and of production;

– human resource management;

– internal actors in the company;

– human, social and institutional environment;

– biophysical environment;

– ethical values goals.

Table 3.3 in the Chapter 3 Appendix describes these nine domains.

As regards the fifteen criteria, they are outlined in Table 3.2. All other things being equal, they give a summary of the different basic principles of the tools, principles, internal and external standards, some of which we have discussed in this section.

Criterion	Description
1. Activity	– relation between the occupation observed and the opportunities offered by the human resources and equipment
2. Internal citizenship	– organization's capacity to encourage internal democracy
3. Local and economic citizenship	– modes of behavior with the actors in the immediate geographic environment wishing to work in the interests of the common or general interest – modes of behavior with the external economic partners wishing to work in the interests of the common interest
4. Competitiveness	– capacity to deal with competition by adapting to changes in the market
5. Conviviality	– capacity to have positive rapports between people (ambience, relationships, etc.)
6. Creativity and esthetics	– capacity for innovation, development of new ideas – capacity to provide visual quality with harmony of shapes and colors
7. Efficacy and efficiency	– ratio between the results obtained and the objectives set – ratio between the results obtained and the means or resources mobilized

Table 3.2. *The 15 criteria for appreciation of the societal balance (source: www.cjdes.org)*

8. Employability and development of competences	– capacity for an employee to find employment elsewhere – capacity to develop individual and collective competences (learning organization, qualifying organization, etc.)
9. Ethics	– capacity to respect the values to which the organization adheres
10. Respect for the environment	– capacity to preserve what it is felt ought to be passed down to future generations
11. Satisfaction	– capacity to cater for the requirements of the consumers and/or the employees in their work
12. Health and Safety	– degree of exposure to danger incurred by the personnel, the customers and third parties
13. Solidarity	– capacity for assistance, support of priority target groups
14. Social and collective utilities	– capacity to provide a good or service that caters for a need which is not, or is inadequately, catered for
15. Viability	– capacity of the organization to ensure its longevity and endurance without external intervention

Table 3.2. (continued) *The 15 criteria for appreciation of the societal balance (source: www.cjdes.org)*

3.4. Conclusion

The economic and environmental aspects are important in a sustainable supply chain management approach. However, it seems that progress for any organization and any supply chain inevitably involves the adherence of its members/actors. Indeed, if the actors adhere to the objectives of sustainable supply chain management, each of its aspects (economic, environmental, social/societal) will dynamize the system in its entirety.

3.5. Appendix

Criterion	Description
Goods/Services – customer relations	Positioning of the product/service and impact assessment Coproduction of the product/service Commercial policy and customer relations
Economic management	Suppliers Recourse to outsourcing or delocalization Investment Financial ethics Placement policy Exploitation of results Management control, auditing and evaluation Means and practices
Anticipation – Innovation – Prospective	Which approach and which objectives? Watchfulness, vigil and prospective Professional participations Research and development budget
Organization of workflow and of production	Professions and competences Transparency of organization Valorization and conditions of use of internal resources Participative organization
Human resource management	Recruitment procedures Welcome Participation – motivation Evolution within the company Search for equity Creation and maintenance of jobs Operational training Non-operational training Working conditions Working conditions–safety Working hours

Table 3.3. *The nine domains of the societal balance (source: www.cjdes.org)*

Internal actors in the company	Valorization of voluntary work Staff participation Financial participation Upward trend Support for employees' civil initiatives Participation in providing/finding accommodation for employees Internal communication directed at the employees
Human, social and institutional environment	Opening of the company's infrastructures to external people Transport policy Involvement of employees in local life Relations of the company's elected authorities and local authorities Contribution to local life Material and technical support to training institutions Support for the economic integration of pupils and students Technical support for creativity and creators Supplier relations Relations with the organization involved in employment management (observatories, employers unions, National Employment Agency, APEC (*Agence pour l'emploi des cadres* – Framework Employment Agency), etc.) Partnership with insertion structures to maintain employment Participation in professional institutions and actions Humanitarian and patronage activities Technology transfer Institutional communication Information Publicity

Table 3.3. (continued) *The nine domains of the societal balance (source: www.cjdes.org)*

Biophysical environment	Respect of the site Construction Noise level Existence of risks Identification of risks People and things exposed to risk Standards Cooperation and mobilization Active policy on raw materials, emissions, etc. Initiatives, pre-emptive measures, responsibilities
Aims – Values – Ethics	Transparency of values Communication of the company's values Respect for legislation Respect for humans Transparency and justification of decisions and practices Organization of powers and of control

Table 3.3. (continued) *The nine domains of the societal balance (source: www.cjdes.org)*

Chapter 4

Sustainable Supply Chain Management Balanced Scorecard

4.1. Introduction

The saying goes that "we can only manage what we can measure". In terms of decision-support measuring tools, one of the most influential is the "dashboard" (or *tableau de bord)*.

A dashboard is a managerial tool which helps effect change in companies by the introduction of "physical indicators, indicators not produced by the organization (unlike accountancy and budgetary figures), environmental indicators or indeed transversal indicators" [CHI 94], (p. 50). However, in order to ensure it is widely used, this dashboard should present a "set of a few (five to ten) indicators [which help] managers to become aware of the status and evolution of the systems that they are controlling, and to identify the trends which will influence these systems on a horizon consistent with the nature of their functions" [BOU 01], (p.397-398).

By combining financial and non-financial indicators, the dashboard facilitates a more efficient management of the activities. Nowadays, it is widely acknowledged to be a potent tool for management and tracking. It is arranged in such a way as to offer a complete view of the processes and an appreciation of the actions justified by a service, by a company, or by a group of actors in a supply chain. The main advantage to this approach lies in its usability as a decision support, facilitating the taking of corrective action if necessary [FER 00].

In the same vein as Germain [GER 05], it is possible to put forward five main traits common to the use of dashboards nowadays – namely:

– the combination of financial and physical indicators to evaluate performance as a whole;

– the presence of guiding indicators centered on the actions currently taking place and results indicators;

– the choice of a small number of indicators;

– the desire to express the strategy on an operational level and draw the connection between the indicators and the company's strategic objectives;

– the emphasis on anticipation and *a posteriori* observation.

4.2. Dashboard and logistics: evolution

The aim of logistics is to deal with the pressure of global competition, where competitive advantage lies largely in one's ability to quickly (and reliably) manage the various flows involved in the supply chain. The trend toward the use of dashboards in logistics highlights the fact that it has moved from being viewed as a dialog tool to being viewed as a diagnostics tool.

Firstly, in the mid-1990s, Tchokogué [TCO 95] suggested the creation of a logistical dashboard as a dialog tool. This dashboard should highlight a certain number of parameters and indicators based on two major constraints: the market requirements and the consistency of flow control. The synoptic presentation of a logistical dashboard begins with two objectives. First, we need to emphasize the following:

– the improved efficiency in terms of consistency and integration in the "financial, economic and organizational domains" [TCO 95], (p.49);

– the improved and enhanced competitive position;

– the return on assets (ROA) and the leverage effect on turnover.

Then, we need to show the value of the distribution of external efficiency (e.g. improved quality, the better management of delivery times) and internal efficiency (e.g. decreased cycle times and the operating costs). The starting point for this discussion is the arguments formulated in the performance pyramid by Cross and Lynch [CRO 89].

Secondly, La Londe and Pohlen [LAL 96] wonder about how to put forward a mechanism combining techniques and cost-management approaches, such as Activity Based Costing or Efficient Consumer Response. To this end, they propose a diagnostics tool – supply chain costing – which is able to consolidate the whole of the supply chain and stimulate change within it. The goal of supply chain costing is to "combine performance measurements with costs" [LAL 96], (p.9). The reasoning is based on the analysis of the processes and activities included in the supply chain. This "clearing out" makes it easier to represent activities as complete costs within and between the organizations. The advantage to a hybrid approach such as this is that it marks costs by outgoing activities, thereby facilitating customer-based, product-based or supply chain-based analyzes.

These two trends fit in with the desire to design dashboards able to align as many interests as possible. In the debate over how to evaluate the performance of a logistical approach, it is possible to list a certain number of basic characteristics required for the proper implementation of a logistics-linked dashboard [SCO 91; CHO 94; LAL 96]. These characteristics are:

– ease of use. In order to evaluate a logistical performance, we need to put in place an easy-to-use measuring system. This system must be the result of multi-dimensional measurements, and able to determine the way in which (and the processes by which) the decisions made by all the actors affect the costs in the supply chain. The role of such a system is:

- to communicate information about the costs incurred – financial and otherwise – between the past and the future,

- to suggest indicators that are able to evaluate the measurement of intangible investments, centered around collective skills, customer satisfaction and satisfaction of other partners,

- to reveal the capacity for adaptation and innovation of each stakeholder involved in the supply chain.

Furthermore, the measuring system must be able to compensate for information-hiding between the different actors and for inapt computer infrastructures:

– a decision-support system. The evaluation of the performance of the logistical strategy by use of a dashboard should lead to a simulation of the flows, the aim of which is to eliminate non added-value activities and, thus, to guide improvements in terms of costs, time and quality. Hence, the goal of such a system is to valorize actions, inter-relations and the good circulation of all forms of flows, and to provide diagnostics and implementation of corrective action;

– multi-actor performance analysis. A logistical dashboard should be able to determine the performance of all the actors who affect the costs in the supply chain. The aim behind this characteristic is to situate the behavior of each firm within the total cost of the supply chain, from the acquisition of the raw material to the moment a finished good and/or service goes on the market.

These three characteristics are naturally also valid in the context of reverse logistics.

4.3. The dashboards currently used in logistics

Two dashboards are usually discussed in a literature review in the field of logistics: the Balanced Scorecard and the SCOR® model.

4.3.1. *The Balanced Scorecard*

The Balanced Scorecard (BSC) is a management tool which highlights the importance of a balance between the short and the long term, between the financial and non-financial criteria, or indeed between major and minor indicators. From an internal point of view, by putting in place "strategic maps" [KAP 01; KAP 06], it affords the opportunity to clearly point out the processes and skills in which the company excels.

On the other side of the Atlantic, in the 1990s, Kaplan and Norton [KAP 92; KAP 96] founded their Balanced Scorecard approach, built around a multi-dimensional aspect. With this tool, the two authors attempt to demonstrate the importance of appraising a company's performance by use of a dashboard centered on strategy and vision. Thus, by way of a cause-and-effect relationship between four axes (finance, customer, internal process, organizational learning and innovation), the BSC aims to offer a multi-criterion analysis of the situation, in order to be able to react as quickly as possible. With this goal in mind, for each of the four axes, Kaplan and Norton [KAP 96] stress the importance of not relying solely on tangible and intangible investments to ensure the company's growth, but also to take account of other components: the capacities of the information systems, the employees' potential, their motivation and their accountability, or indeed aligning the objectives between those of the company and those of its internal actors.

The model is constructed around four axes which are considered to be of crucial importance:

– the finance axis serves as a conductive link to the indicators from the other axes of the BSC. The objective is for each indicator selected to be part of a string of cause-and-effect relations, the ultimate aim of which is to improve financial performance;

– the customer axis identifies the sectors of the market in which the company wishes to achieve a position: those which will generate enough revenue to realize its financial objectives. This is a key indicator for a company wishing to become number one in its field of products and services, and for the groups of customers it is targeting;

– the internal processes axis is intended to improve all the processes, from the initial innovation to the after-sales service. The objectives and indicators here are defined on the basis of explicit strategies, aimed at fulfilling the expectations of the shareholders and the target customers. This sequential approach usually reveals new processes in which the company should excel at in the medium term;

– the organizational learning and innovation axis represents the means, resources and factors needed in order to achieve the objectives established on the "finance", "customer" and "internal processes" axes, in the knowledge that the latter identify the domains in which the company should distinguish itself in order to improve its performance.

For as long as it has been around, however, the BSC has been the target of no small amount of criticism. The four measuring axes are not considered to be appropriate for all structures, or for every situation. For instance, occasionally it would be helpful to add: the suppliers, the community: the state, local collectives, etc. [ATK 97; EDV 97; BUR 00; FER 00; SUP 02; MOR 02] or the environment [HOC 01].

Similarly, certain aspects of the model ought to be "subdivided" and made more specific. Thus, and particularly as regards the "organizational learning and innovation" category, which is perceived as being too broad, it would be desirable to make more of a point about the contribution of the company's employees and innovation by research and development [MAL 03].

Finally, certain authors stress that in practice, the use of the model is based primarily on three dimensions of the BSC: finance, the customer and the internal process, largely leaving the fourth dimension aside – i.e. organizational learning and innovation [SPE 03]. Thus, [SPE 03] also harbors the idea that the BSC is far from a "clearly-defined concept" (p.362) because in practice, each company "personalizes" it, adapting it to its own requirements. A study put to 42 companies in Germany which adopted a BSC shows three different types of BSC: type I, which is a multi-dimensional structure that combines financial and non-financial strategic measurements; type II which includes type I plus a description of the cause-and-effect relations used; and type III, which combines type II plus an implementation of the strategy by definition of objectives, plans of action, results and bonuses.

Certain criticisms leveled at the BSC are even sharper: the construction would not work with a rigorous approach. Norreklit [NOR 00] points out that there are not always causal links between the different measuring axes, and that it constitutes a model specific to "top management" only, which has no place in organizational operations. In his negative assessment, Norreklit even stated [NOR 03] that the success of the BSC was attributable more to the "rhetorical" use of the meaning attached to it by Kaplan and Norton than to its content. This critical discourse is partly echoed in French discourse, when a comparison is drawn between the BSC and the French *tableau de bord*. In the eyes of Chiappelo and Lebas [CHI 96], the BSC was essentially inspired by the French tool; for Bourguignon *et al.* [BOU 04], it is a "fashionable" tool with no tradition; and for Bessire and Baker [BES 05], its success is largely down to the help of consultancy firms in promulgating it.

Although it has been criticized and is indeed criticable, the BSC nevertheless presents an opportunity to ask questions about the advantage and importance of multi-dimensional guidance tools to make an industrial company more efficient and effective [HOQ 00; MAL 03; SPE 03; MIC 05]; and it is studied in order to deal with the issues encountered in logistics and SCM [LIB 98; BRE 00; MOR 00; MOR 02].

4.3.2. *SCOR®: Supply Chain Operations Reference model*

The question of development of a set of metrics relating to SCM has also, for many years, been the subject of discussions between companies, academic institutions and the management consultancy firm "Pittiglio, Rabin, Todd & McGrath" (PRTM). Introduced in 1996, the SCOR® (Supply Chain Operations Reference®) model was devised by the Supply Chain Council (SCC[1]), which was formed by the consultants PRTM[2], AMR Research[3] and, initially, 69 enterprises in differing economic sectors.

The SCOR® model can be defined as a guidance tool, which originates with a reflection on the analysis of five key processes of management, enumerated as:

– plan;

– source;

– make;

– deliver;

– return.

The five key processes of management refer, first of all, to two categories of costs linked to customer relations and internal operations. These are then subdivided, respectively, into three and then two main dimensions [STE 96; PIT 99]:

– customer relations, including:

- reliability, which is the capacity to correctly deliver the right product to the right place, in the desired timeframe, in the required packaging conditions, in a sufficient quantity, with the appropriate documentation and to the right customer,

- reactivity, which involves the goal of quickly providing each customer with the products,

---

1 The Supply Chain Council was formed in 1996-1997 on the initiative of PRTM, AMR Research, Procter & Gamble, Texas Instruments and other organizations. Its role is to promote and valorize the use of the SCOR® model (www.supply-chain.org).

2 PRTM (Pittiglio, Rabin, Todd & McGrath) is a consultancy firm founded in 1976, with the goal of providing technological support to many international companies (www.prtm.com).

3 AMR (Advanced Manufacturing Research), created in 1986, is a research group specializing in the programming of systems to measure the performance of process-based management tactics (www.amresearch.com).

- flexibility, which is the ability to adapt to changes in the environment;

– internal operations, including:

- the costs, which are the financial reflection of the operationalization of the supply chain,

- the efficiency of asset management, which expresses the satisfaction of the demand by efficient management of the fixed and variable assets of the organization.

Yet in current logistical practices, at least in France, the Supply Chain Council's SCOR® model does not really tend to be used. For instance, Morana [MOR 08] highlights the costs of purchase and the implementation of this tool, with over 150 indicators, as being barriers to its use in company policies in France, particularly for small and medium enterprises (SMEs). The idea of a combination between the Balanced Scorecard model and the logistics model appears to be more successful. The actual design of the tool, enhancing the axes of action by enumerating financial and non-financial indicators, would cater more accurately to the requirements of the market and of the customers.

However, as regards the practice of sustainable supply chain management, one wonders whether it might not be of more interest to construct a tool founded on the idea of the Balanced Scorecard, highlighting the economic, environmental and social/societal axes. This division could – all other things being equal – facilitate or reconcile the construction of a guidance tool of this type. Thus, the economic axis would focus more on indicators relating to costs, quality and delay: three generic elements in the appraisal of a supply chain; the environmental axis would relate to eco-design, waste-management, reverse logistics and remanufacturing; and the social/societal axis would look at the role and weight of the different stakeholders in the supply chain – both internal and external. We advocate the graphic representation shown in Figure 4.1.

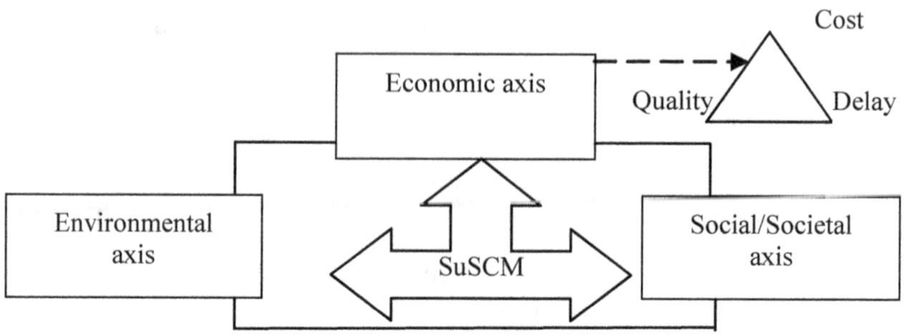

Figure 4.1. *Sustainable Supply Chain Management Balanced Scorecard (SuSCMBSC)*

4.4. The indicators used in Sustainable Supply Chain Management Balanced Scorecard

The aim of this section is to list and discuss the indicators relating to transport and logistics. It goes without saying that these lists are far from exhaustive, as the question of the list of indicators for assessing transport and logistics is still a matter for some considerable debate.

4.4.1. *Economic indicators*

The economic indicators reflect the economic aspect of SuSCM. In fact, using the reasoning put forward in Chapter 1 of this book, we find indicators relating to costs, quality and delays. More specifically, the indicators relate to upstream, production and downstream logistics.

A few suggestions for indicators are set out below.

4.4.1.1. *Strategic and operational indicators of SCM*

Based on the list given by Gunasekaran *et al.* [GUN 01] and on the addition of indicators after a pre-test, the results of an analysis run on a number of branches of a hi-tech multinational enable us to highlight two categories of dashboards relating to SCM [MOR 02]. Table 4.1 enumerates the indicators deemed to be highly important by the surveyed population, which we class as strategic indicators. Table 4.2

lists the indicators considered to be less important, which we classify as operational indicators.

*** Class 1: Cycle of production of the product/service, from the initial supplier to the final customer**		
Financial – nmbr: 3	**Non-financial – nmbr: 9**	**Financial / Non-financial – nmbr: 1**
– Total inventory as incoming stock level; – Total inventory as work in progress; – Total inventory as finished goods in transit	– Delivery lead time; – Quality of information; – Responsiveness to urgent deliveries; – Flexibility of service systems to meet particular customer needs; – Delivery performance; – Accuracy of forecast; – Total supply chain cycle time; – Master production planning flexibility; – Order lead time	– Buyer-supplier partnership level
*** Class 2: Management of quality as a whole: materials, machines, information**		
Financial – nmbr: 2	**Non-financial – nmbr: 3**	**Financial / Non-financial – nmbr: 0**
– Total inventory as scrap; – Rate of return on investment (ROI)	– Capacity utilization; – Level of customer perceived value of product; – Quality of delivered goods	
*** Class 3: Delivery of the product**		
Financial – nmbr: 0	**Non-financial – nmbr: 2**	**Financial / Non-financial – nmbr: 0**
	– Customer relationship depth; – Supplier lead time against industry norm	

Table 4.1. *Strategic indicators of SCM (source: adapted from [MOR 02])*

<table>
<tr><th colspan="3">* Class 1: Cycle of production of the product/service, from the initial supplier to the final customer</th></tr>
<tr><th>Financial – nmbr: 0</th><th>Non-financial – nmbr: 4</th><th>Financial / Non-financial – nmbr: 0</th></tr>
<tr><td></td><td>– Planned process cycle time;
– Effectiveness of distribution planning schedule;
– Information propagation cycle time;
– Product throughput efficiency</td><td></td></tr>
<tr><th colspan="3">* Class 2: Management of quality as a whole: materials, machines, information</th></tr>
<tr><th>Financial – nmbr: 0</th><th>Non-financial – nmbr: 3</th><th>Financial / Non-financial – nmbr: 2</th></tr>
<tr><td></td><td>– Information carrying cost;
– Achievement of defect-free deliveries;
– Level of supplier's defect-free deliveries</td><td>– Risk liability;
– Supplier rejection rate</td></tr>
<tr><th colspan="3">* Class 3: Delivery of the product</th></tr>
<tr><th>Financial – nmbr: 0</th><th>Non-financial – nmbr: 3</th><th>Financial / Non-financial – nmbr: 0</th></tr>
<tr><td></td><td>– Driver reliability for performance;
– Efficiency of purchase order cycle time;
– Purchase order cycle time</td><td></td></tr>
<tr><th colspan="3">* Class 4: Administrative management</th></tr>
<tr><th>Financial – nmbr: 1</th><th>Non-financial – nmbr: 8</th><th>Financial / Non-financial – nmbr: 1</th></tr>
<tr><td>– Net profit to productivity ratio</td><td>– Order entry methods;
– Effectiveness of delivery invoice methods;
– Supplier's booking in procedures;
– Supplier ability to respond to quality problem;
– Supplier assistance in solving technical problems;
– Range of products and services;
– Quality of delivery documentation;
– Adherence to master production schedule</td><td>– Customer query time</td></tr>
</table>

Table 4.2. *Operational indicators of SCM (source: adapted from [MOR 02])*

4.4.1.2. *The indicators from the SCOR® model*

Here, we shall focus only on the level-1 indicators from the SCOR® model.

4.4.1.2.1. The indicator "delivery performance"

In general, we can state that a delivery can be understood as the physical handover of a good to the customer or to the customer's representative, who accepts it. The improvement of the delivery performance is synonymous with making deliveries on the dates promised to the customers.

A few examples of indicators:

– rate of on-time delivery (conformance with the agreed time-delays);

– rate of delivery in the correct quantity, which can be divided into partial and total delivery;

– rate of complaint, which can be divided into various types of complaints (the documentation not included, a manufacturing fault, a missing part, etc.).

4.4.1.2.2. The indicator "order entry"

An order is a manifestation of a desire which connects a customer with a supplier, with the intention of making available a good and/or service that helps cater for a need and/or to realize a project. Hence, the order relates both to an object (the delivered good and/or service) and to a procedure (all of the stages, which run from the initiation by the order issuer, through manufacturing of the product, to its final reception). Here, the starting point must (or should) be taken at the "order confirmed" – i.e. after the period afforded to the customer to retract the order if they so desire.

A few examples of indicators:

– for the realization time: the average time taken to process an order line;

– for perfect realization: the number of rejects, the number of certified quality orders;

– for the rate of availability: the ratio of the response delay to parts in stock, which can be divided into national and international categories.

4.4.1.2.3. The indicator “response time of supply chain”

According to the definition given by the SCOR® model, the overall supply chain should be understood as the set of all the informational, financial and physical flows which lead from the suppliers’ suppliers to the customers’ customers.

The supply chain response time may be viewed, for instance, through the lens of the total time taken from the extraction of the raw material to the delivery of the product. It may also refer to the time taken to react to evolutions of the market in order to minimize this production time as far as possible, by increasing the number of new customers, the percentage of registered patents or indeed the number of partnerships for “enhanced” monitoring.

4.4.1.2.4. The indicator “flexibility of production”

Flexibility of production can be analyzed as the ability to quickly alter the characteristics of a good so as to adapt it to the peculiarities of the order. It may thus be taken to refer to the mechanism by which we move from a push-flow to a pull-flow policy.

An example of an indicator:

– number of specific supplies (the number of orders personalized).

4.4.1.2.5. The indicator “total cost of supply chain management”

If we accept the definition of the supply chain as given above, the total cost of supply chain management can be identified by adding up the different costs – upstream and downstream – relating to transport, supply and manufacture.

4.4.1.2.6. The indicators "cost of guarantee" and "cost of recycling"[4]

These two costs can refer to all the elements included in the concept of green operations (see Chapter 2). Consequently, the "cost of guarantee" is comparable to the costs of maintenance and repair of returned goods (costs of repair, detached parts, fault diagnosis) and the "cost of recycling" to the cost of waste management (destruction).

4.4.1.2.7. The indicator "cycle of liquidity turnover"

This level of cost may, for instance, be examined through the lens of RONA (Return On Net Assets) which "represents the time needed to recover the capital invested in the company" [CAD 95] (p.433). The more distant the connection between each element making up the financial cycle of the RONA, the harder it becomes to make the strategic decisions. The RONA, or economic profitability, corresponds (in %) to the ratio of the operating surplus to the economic assets. This ratio causes a financial cycle which loops the costs of purchasing, warehousing, production, storage, sales, delivery, billing and debt collection.

4.4.1.2.8. The indicator "added-value productivity"

Here, we propose to examine "added-value productivity" by way of two elements.

To begin with, consider the hourly productivity index, which refers to the change of the added value in costs and the variation in the volume of hours worked (in sets of full hours).

Then, analyze the intermediary balances carried forward in the profit and loss account, which involves calculating the gross added value, with the aim of measuring the company's activity which contributes to the achievement of the result. Another indicator which is also interesting in terms of the balances carried forward is the gross operating surplus (GOS), which is more meaningful in the sense that it takes account of the impact of salary costs in the analysis.

---

4 In the logic of SuSCMBSC, these two indicators need to be switched around on the environmental axis.

4.4.1.2.9. The indicator "stock rotation"[5]

This indicator is one of the most commonly-used operational indicators. It entails determining the number of times stock is renewed during a given period. It can be calculated for stocks of raw materials, goods, intermediary products, semi-finished products, and finished products.

4.4.1.2.10. The indicator "stock inventory days"

This indicator determines the number of days that stocks are kept at the company. It also reveals idle inventory.

4.4.1.3. *Process performance indicators*

In the literature published in the area of logistical performance, we find indicators relating to the processes. For instance, Petitqueux [PET 06] proposes six performance indicators for the processes that are part of a supply chain (see Table 4.3).

Indicator	Description
Industrial Efficiency (IE)	Efficiency of the processes and revelation of wastage
Overall Equipment Effectiveness (OEE)	Measurement of automated processes and demonstration of proper use of industrial means
Process Lead Time (PLT)	Time taken between the initial operation and delivery to the store (expressed in hours)
Inventory Coverage (IC)	Management of stocks and partial products in relation to the volumes produced (expressed in days)
Production Changeover Times (PCT)	Downtime for the production line during changeover between two products/series
On-Time Delivery (OTD)	Respect of commitments regarding the delivery of the products

Table 4.3. *Process performance indicators (source: [PET 06])*

5 The indicators "stock rotation" and "stock inventory days" are part of an operational thought process. They apply, in particular, to the context of warehouse management. On this point, Mocellin [MOC 06] suggests a number of indicators relating to this type of evaluation.

4.4.1.4. *Urban logistics indicators*

In terms of urban logistics, we look at the management of the auto fleet in the last link in the supply chain. In that sense, the work done at the *Laboratoire d'économie des transports*[6] on urban transport of goods [PAT 09] offers a descriptive state-of-the-art on such measurements (see Table 4.4).

Name and description of the indicator	Unit in which the indicator is measured
Ratio of deliveries to collections per use in a particular sector of activity	Number of deliveries/ collections per week per employee
Density of deliveries/collections in an area	Number of deliveries/collections per week per km^2
Intensity of deliveries/collections in a sector of activity in an area	Number of deliveries/collections per week
Length of time spent double-parked during deliveries/collections to deliver or collect in a sector of activity in an area	Number of hours
Average distance covered to deliver or collect in a sector of activity with a type of vehicle	Number of km
Length of walk from loading bay to site of delivery	Number of km
Average distance needed to deliver/collect	Km for deliveries/collections
Total distance covered in urban areas by goods vehicles	Vehicles per km per week
Average duration of a delivery/collection (per sector of activity, type of vehicle, mode of management)	Minutes per delivery
Average speed per round (including or excluding stops for actual deliveries)	Km/h
Weight transported per kilometer of transport depending on the sector of activity, type of vehicle, mode of management	Tons (or kg) per km

Table 4.4. *The indicators produced by the "transport of goods in urban areas – economic aspect" (source: [PAT 09])*

6 The *Laboratoire d'économie des transports* (LET) is a "research laboratory specializing in the transport economy and land use" (source: www.let.fr/fr/let_en_bref/index.html).

As already stated, these lists of economic indicators are far from exhaustive. There are many works devoted to this issue in the existing body of literature. However, we believe this section has given a relatively broad overview of the current fields of discussion about the economic performance of supply chain management.

4.4.2. *Environmental indicators*

As yet, the practices of environmental indicators in logistics have not been the subject of much discussion. The indicators "cost of guarantee" and "cost of recycling" from the SCOR® model have a place in our discussion of them here.

In terms of transport management and urban logistics, we can mention [PAT 09]: energy consumption, pollution, the greenhouse effect depending on the area, the vehicle, the activity, the mode of management (total or per movement):

– in tons of diesel;

– in kg of pollutant (Nox, SO_2, CO, PM, etc.);

– in kg of CO_2.

The intention of this indicator is to quantify the "impact of movement of goods on energy consumption, local and global problems and greenhouse gas (GHG) emissions" [PAT 09], (p.33).

On the subject of eco-design, we can look at the recyclability of the manufactured product.

More generally, we can turn to Depoers *et al.* [DEP 03] who set out a list of indicators for measuring companies' performances in terms of sustainability, on the basis of varied sources such as the World Business Council on Sustainable Development, the Association of British Insurers, the Global Reporting Initiative[7], an ethical funding

7 Remember, the Global Reporting Initiative was set up in 1997, and is the architecture listing the principles and indicators that any organization can use in an

provider (Branics), specific Websites (UNAIDS, ADEME, etc.) and the companies themselves. In view of the elements cited in green supply chain management, we believe these indicators have a leading role to play here.

Thus, of the seven objectives cited, the first three relate to environmental measurements. For each objective, themes are suggested, coupled with various generic indicators:

– objective 1: overall environmental impact:

- global atmospheric threats: (1) GHG emissions[8], (2) emissions of substances that deplete the ozone layer, (3) emissions of substances contributing to acidification, (4) average energy consumption (automobile),

- marine pollution: average energy consumption (maritime transport),

- biodiversity: (1) forest management – diversity of fuels, (2) forest management – PEFC (Pan-European Forest Certification scheme) and FSC (Forest Stewardship Council scheme), (3) grain variety,

- risk management: degree of ISO 14001 certification;

– objective 2: management of resources whilst preserving the local environment:

- resource management: (1) water intensity, (2) energy intensity, (3) raw materials intensity, (4) raw materials recycling, (5) product energy consumption,

- prevention at the design stage: eco-design,

- local pollution – water pollution: (1) content in toxins and metals, (2) pollutant content, (3) biodegradability index,

---

economic, environmental and social context (www.globalreporting.org/AboutGRI/WhatIsGRI/).

8 The methods for measuring greenhouse gas emissions are presented in section 2.5.4.

- local pollution – soil pollution: (1) soil pollution – hazardous waste, (2) soil pollution – rehabilitation of sites;

– objective 3: access to vital products: water and foodstuffs:

- local pollution – water pollution: (1) content in toxins and metals, (2) pollutant content, (3) biodegradability index,

- local pollution – soil pollution: (1) soil pollution – hazardous waste, (2) soil pollution – rehabilitation of sites, (3) number of hectares lost.

In the same vein as the work of Depoers *et al.* (2003), we note the study carried out by the *Observatoire* on corporate social responsibility [ORS 03] on a comparison of indicators on sustainable development published by Cer2D (*Centre d'études et de recherches sur le développement durable* – Center for R&D on Sustainable Development), the law on new economic regulations (NRE) and the Global Reporting Initiative, which tends toward the same type of generic indicator statements.

Finally, to conclude this section, a concordance table of the environmental indicators between the NRE law and the GRI [DEP 04] can also be used in a logistics context (see Table 4.5).

Environmental indicators from the NRE/GRI	
NRE, article 2 of Decree n° 2002-221 for the application of article L225-102-1 of the Code of Commerce	**Global Reporting Initiative (environmental section)**
Consumption of water resources	EN5. Total consumption of water EN20. Sources of water supply and connected ecosystems/habitats greatly affected by water consumption EN21. Sampling in the phreatic zone and at the surface, regarding percentage available in supply sources EN22. Total volume of water recycled and reused

Table 4.5. *NRE/GRI environmental indicators (source: [DEP 04])*

Consumption of raw materials	EN1. Total consumption of raw materials per type EN4. Percentage of materials consumed which are waste products (recycled or otherwise) from sources outside the company
Energy consumption. Measures to improve energy efficiency and use of renewable energies	EN3. Direct energy consumption, distributed by primary sources EN4. Indirect energy consumption EN17. Initiatives to exploit renewable energy sources and improve energy yield EN18. Energy footprint of products EN19. Other forms of indirect energy consumption
Condition of soil use	EN16. Site and surface of terrains held, rented or managed in biodiversity-rich habitats EN23. Total surface area of terrains held, rented or managed for activities of production or extraction EN24. Total impermeable surface as a percentage of terrains bought or rented EN28. Number of endangered species (IUCN Red List) in areas affected by the company's activities EN29. Production units within or around protected or sensitive zones

Table 4.5. (continued) *NRE/GRI environmental indicators (source: [DEP 04])*

Emissions into the air, water or soil, drastically affecting the environment (enacting decree, 30 April 2002)	EN7. Description of the main impacts on biodiversity caused by the company's activities and/or products or services EN8. Greenhouse gas emissions in equivalent tons of CO_2 EN9. Use and emissions of substances which deplete the ozone layer in equivalent tons of CFC-11 EN10. Emissions of NOx, SOx and other atmospheric pollutants EN12. Significant emissions into the water, per type EN13. Significant accidental leakage of chemical products, oils and fuels (number and total volume) EN14. Environmental impacts of the products and services EN25. Impacts of activities on protected and sensitive zones EN26. Modifications of natural habitats due to activities and percentage of terrains protected or restored EN30. Other relevant indirect GHG emissions EN32. Sources of water supply and connected ecosystems/habitats greatly affected by emissions and seepages EN34. Environmental impacts of means of transport
Sonic and olfactory nuisances	
Measures to limit the damage to biological equilibrium, natural environments and protected animal and plant species	EN27. Objectives, program and provisions for protection and restoration of the indigenous ecosystems and species in degraded zones
Approaches for environmental evaluation or certification	

Table 4.5. (continued) *NRE/GRI environmental indicators (source: [DEP 04])*

Waste	EN11. Total quantity of waste per type and destination EN15. Percentage mass of products sold that can be recovered and reused at end of life, and percentage that actually *is* recovered EN31. Production, transport, import or export of all waste deemed "hazardous" under the terms of the Basel convention
Amount paid out in indemnities in court cases over environmental matters	EN16. Case and penalties for non-respect of environmental law
Amount of provisions and guarantees for environmental risks	
Measures taken, if need be, to ensure the company's activity does not violate environmental laws	
Environmental expenditure committed	EN35. Total environmental expenditure per type
Existence of internal environmental management services, training and employee information. Objectives assigned to affiliates abroad (1st and 6th points of the Decree)	
	EN33. Suppliers' environmental performances

Table 4.5. (continued) *NRE/GRI environmental indicators (source: [DEP 04])*

4.4.3. *Social/societal indicators*

To date, little work has been done in terms of enumerating social/societal indicators in supply chain management. In fact, we propose to present a list of such indicators published by the IFAC – International Federation of Accountants [IFA 98], taken up again by Burlaud [BUR 00] with a view to management of the intellectual capital in an organization which, we believe, fits in with and can be subdivided in to thinking about supply chain management (see Table 4.6).

1. Human capital indicators
Reputation of the company's employees with head-hunters
Number of years experience in the profession
Percentage of employees with less than two years experience
Satisfaction of employees
Proportion of employees putting forward new ideas (which are implemented)
Added value per employee
Added value per dollar of salary paid

2. Organizational capital indicators
Number of patents
Turnover per R&D project
Cost of patent renewal
Number of connections from individual computers to the database
Number of consultations of the database
Number of contributions to the database
Number of updates of the database
Number of connections and connection time to the computer system
Cost of the computer system per dollar of turnover
Rate of satisfaction of services provided by the computer system
Number of new ideas presented over number of new ideas applied
Number of new products
Number of new products per number of employees
Number of multi-functional project teams
Share of turnover generated by new products
Trend over five years in the lifecycle of the products
Average duration of design and development of new products
Values of the new ideas (savings or revenue)

3. Relational capital indicators
Increase in volume of revenue
Percentage of sales made to faithful customers
Brand fidelity
Customer satisfaction
Customer complaints
Products returned as a percentage of sales
Number of agreements reached with customers or suppliers and amount of business thus generated

Table 4.6. *Measurements for management of intellectual capital (source: [IFA 98], p. 11, reproduced in [BUR 00])*

As regards ANACT (*Agence nationale pour l'amélioration des conditions de travail*[9] – France's national agency for improving working conditions), they put forward three types of indicators which can be highlighted in the fields of transport and logistics. We find indicators relating to:

– workplace accidents, with:

- the frequency,

- the seriousness,

- the nature of the injuries: type and position on the body, origin (during which phase of the task),

– workplace illnesses;

– absenteeism:

- rate of absenteeism,

- % of stoppage of work for more than *x* days.

Regarding works relating to the measurement of sustainable social/societal performance, it is possible to cite – as cited above in the list of environmental indicators – the list of objectives/themes/indicators put forward by Depoers *et al.* [DEP 03] in terms of measurement of sustainable development. Here, we find the fourth, fifth and sixth objectives of their summary work:[10]

9 www.anact.fr/.

10 The seventh and last objective on the list given in [DEP 03] is associated with shared sustainable economic development (solidarity towards developing countries) which looks at:

– the local creation of added value in developing countries, with the share of turnover in developing countries on the one hand, and the added value created and reinvested in developing countries on the other;

– the deployment of civil society policies, with the rate of deployment of policies of international solidarity.

This indicator fits in with the idea of "socially-responsible consumption" as presented in section 3.3.3.2. Indeed, we speak of "trade which is fair to small producers in developing countries". In itself, were it to be adopted, a breakdown over the whole of a national territory would also be appropriate.

– 4th objective: contribution of industry to public health:

- prevention and treatment of epidemics: lowering of prices and donations of essential medical supplies,

- direct control of health-related expenses: provision of generic medicines,

- indirect control of health-related expenses: (1) seriousness of workplace accidents, (2) frequency of workplace accidents, (3) health-and-safety training in the workplace, (4) workplace illnesses, (5) exposure to carcinogenic risk,

- food safety: (1) degree of HACCP (Hazard Analysis Critical Control Points) certification, (2) rate of supply in sustainable farming, (3) conformity with food safety regulations;

– 5th objective: training, education and cultural development:

- training of employees and subcontractors in developing countries:[11] (1) effort invested in training employees, (2) effort invested in training subcontractors,

- access facilitated to new information and communication technology (ICT): Internet access for employees;

– 6th objective: the design of products and services oriented toward social demand:

- promotion or support of territories: policy of promotion or support of territories;

- promotion of local products: elaboration of local products, distribution and promotion of local products.

4.5. Conclusion

In the context of sustainable supply chain management, it is important to have a dashboard which has common indicators. This tool should be explicit both to people within the organization and to

11 On this point, this theme can also be adapted in the country in its own right.

the whole of the supply chain. The most difficult thing here is to find the key indicators, in reduced number (remember that Bouquin [BOU 01] favors a management tool with at most ten indicators) and which are apt to respond to the three desired economic, environmental and social/societal dimensions.

General Conclusion

Businesses face not only economic but also environmental and social/societal challenges if they wish to stay afloat, in the national and the international economy. Indeed, on the one hand, if they are not able to meet these three challenges, they will, in the short term, be doomed to fail – doomed by the end consumers, who will gradually turn away from their product and/or service, in favor of those which are not only surer but cleaner. In addition, these companies will ultimately be condemned by the public authorities, who will not always be able to offer and/or accept pollution quotas for an economic good which, in the long run, turns out to be more destructive…

On the other hand, companies which show themselves to be pioneering in their missions and objectives, and which are able to formulate their production and control tools so as to rise to these three challenges, can be assured of gaining in terms of image, quality and finance.

The issue of sustainable supply chain management is important at present, when the Earth itself is beginning to show less and less willingness to "accept" the industrial misconduct which often occurs in spite of environmental policies. Sustainable construction, production and development are becoming crucial. Certainly, this may appear to be utopian thinking in a world where the quest for benefits remains a key element (we have to keep the shareholders and bankers happy, as they are the main financial investors), but more and more states are becoming aware of the pressing need to change the way in which we consume, and are beginning to offer aid to steer us toward production which is more profitable – in the broadest sense of the term – to each stakeholder involved in a company – or indeed in a nation.

The aim of this book was to establish a state of the art – as exhaustive as possible – on the tools/methods/methodologies/audits, etc. to facilitate a sustainable approach to logistics. It goes without saying that the implementation of a sustainable supply chain management approach is not something that can be done overnight. It must be part of a genuine desire for real change – change which must first come from the directors of a company, and then be enacted by each member of each company within a supply chain. In other words, it is a question of gaining a view of the circumstantial position of each company, improving the workings of the mechanism and particularly coordinating each cluster encountered. This is true for every internal service, but also for every company within a supply chain, and for all the "secondary" stakeholders in that chain (NGOs, the State, the media, etc.).

Sustainable supply chain management, beyond the design of standards and tools, is a strategic approach which the company has to fully embrace. Firstly, if it does not aim for an economic, environmental and social/societal systemic logic, it cannot hope to survive in a future where financial resources, natural resources and… skills! ... are in increasingly short supply. Secondly, it is by demonstrating more innovating R&D and by being able to (re)use that which is and has been, that the company will be able to progress and stay center-stage.

> "When each and every individual applies himself to progress, then Humankind will be making progress."
>
> Private diaries of Charles Baudelaire (1821-1867)

Bibliography

[ADE 99] ADEME, *Conception de produits et environnement 90 exemples d'éco conception*, ADEME Editions, Paris, 1999.

[ADE 06] ADEME, *Eco-conception en actions*, ADEME Editions, 2nd edition, Paris, www.ademe.org, 2006.

[AHM 02] AHMED S., SCHROEDER R.G., "Refining the Product-Process Matrix", *International Journal of Operations & Production Management*, vol. 22, no. 1, p. 103-124, 2002.

[AMM 05] AMMENBERG J., SUNDIN E., "Product in environmental management systems: drivers, barriers and experiences", *Journal of Cleaner Production*, vol. 13, no. 4, p. 405-415, 2005.

[ARB 03] ARBAOUI S., HAURAT A., OQUENDO F., THEROUDE F, VERJUS H., "Languages and mechanisms for software processes and manufacturing enterprise processes: similarities and differences", *Proceedings of 5th International Conference on Enterprise Information Systems*, Angers, France, September 2003.

[ARB 07] ARBAOUI S., MORANA, J., "Réflexions sur l'élaboration d'un guide EDI pour les PME", *Revue Sciences de Gestion*, no. 222, p. 131-140, 2007.

[ART 03] ARTOUS A., L'évolution du métier de prestataire logistique et l'expansion du marché européen de la logistique, Summary report, SES, July/August 2003.

[ATK 97] ATKINSON A., WATERHOUSE J., WELLS R., "A stakeholder approach to strategic measurement", *Sloan Management Review*, vol. 38, no. 3, p. 25-37, 1997.

[AUR 97a] AURIFEILLE J.M., "Marketing et logistique: du soutien tactique à la collaboration stratégique", in AURIFEILLE J.M. *et al.* (ed.), *Management logistique: une approche transversale*, Litec, Paris, p. 5-32, 1997.

[AUR 97b] AURIFEILLE J.M., COLIN J., FABBE-COSTES N., JAFFEUX C., PACHÉ G., *Management logistique : une approche transversale*, Litec, Paris, 1997.

[BAR 00] BARDOU L., "Soutien Logistique Intégré", *Technique de l'Ingénieur*, 2000.

[BAR 98] BARRIOL N., "Le marketing achat : vingt-cinq ans d'histoire", *Cahier de Recherche*,no. 98-05, ESA-CERAG, 1998.

[BOU 06] BOUDOUIN D., *Guide méthodologique: les espaces logistiques urbains*, La Documentation française, Paris, 2006.

[BEA 00] BEAULIEU M., "Définir et maîtriser la complexité des réseaux de logistique à rebours", *Les Troisièmes Rencontres Internationales de la Recherche en Logistique (RIRL)*, Trois-Rivières, Quebec, 9, 10 and 11 May 2000.

[BEA 08] BEAULIEU M., LANDRY S., "La matrice produit-procédé: de l'usine à la chaîne logistique", *Logistique & Management*, vol. 16, no. 1, p. 23-29, 2008.

[BEA 99a] BEAULIEU M., MARTIN R., LANDRY S., "Logistique à rebours-: un portrait nord-américain", *Logistique & Management*, vol. 7, no. 2, p. 5-14, 1999a.

[BEA 99b] BEAULIEU M., MARTIN R., LANDRY S., "Logistique à rebours: synthèse de la littérature et typologie", *Cahier de recherche*, no. 99-01, HEC Montreal, 1999b.

[BER 09] BERNEMAN C., LANOIE P., PLOUFFE S., VERNIER M.F., "L'éco-conception: quels retours économiques pour l'entreprise", *Cahier de Recherche*, no. IEA-09-03, HEC Montreal, 2009.

[BES 05] BESSIRE D., BAKER C., "The French tableau de bord and the American balanced scorecard", *Critical Perspectives on Accounting*, vol. 6, p. 645-664, 2005.

[BIG 95] BIGRAS Y., HALLEY A., GÉLINAS R., "Juste-à-temps et transport dans le marché nord-américain", *Actes de la 30e Conférence annuelle du Canadian Transportation Research Forum*, Aylmer, Canada, 15-17 May 1995.

[BLA 08] BLANQUART C., CARBONNE V., "La place du transport dans les organisations logistiques durables", *Les Cahiers Scientifiques du Transport*, no. 54, p. 11-24, 2008.

[BOI 04] BOISSIN J.P., CASTAGNOS J.C., GUIEU G., LIVOLSI L., "Emergence et structuration de la pensée scientifique en logistique. Une étude bibliométrique des quatre premières RIRL (1995-2002)", *5es Rencontres Internationales de Recherche en Logistique*, Fortaleza, Brazil, 2004.

[BOI 06] BOISSIN J.P., CASTAGNOS J.C., GUIEU G., LIVOLSI L., "Structuration de la pensée scientifique en logistique: une étude bibliométrique des cinq premières RIRL (1995-2004)", *Logistique & Management*, vol. 14, no. 1, p. 59-70, 2006.

[BOI 04] BOIRAL O., "Mettre en œuvre ISO 14001: de la quête de légitimité à l'émergence d'un "mythe rationnel", *13e Conférence de l'AIMS*,Vallée de Seine, France, 2-4 June 2004.

[BON 99] BONET D., Conflit et coopération dans le canal de distribution: l'analyse du discours des acteurs comme révélateur des comportements stratégiques, Doctoral Thesis, University of the Mediterranean Aix-Marseille II, June 1999.

[BON 05] BONET D., PACHÉ G., "A new approach for understanding hindrances to collaborative practices in the logistics channel", *International Journal of Retail & Distribution Management*, vol. 33, no. 8, p. 583-596, 2005.

[BOU 01] BOUQUIN H., *Le contrôle de gestion*, 5th edition, Presses Universitaires de France, Paris, 2001.

[BOU 04] BOURGUIGNON A., MALLERET V., NORREKLIT H., "The American balanced scorecard versus the French tableau de bord: the ideological dimension", *Management Accounting Research*, vol. 15, p. 107-134, 2004.

[BOW 99] BOWERSOX D., CLOSS D., STANK T., *21stCentury Logistics: Making Supply Chain Integration a Reality*, Council of Logistics Management, Oak Brook, 1999.

[BRE 00] BREWER P., SPEH T., "Using the balanced scorecard to measure supply chain management", *Journal of Business Logistics*, vol. 21, no. 1, p. 75-93, 2000.

[BRE 97] BREZET J.C., VANHEMEL C., *Ecodesign – a Promising Approach to Sustainable Production and Consumption*, UNEP, United Nations Publication, Paris, 1997.

[BRI 05] BROWNE M., SWEET M., WOODBURN A., ALLEN J., Urban freight consolidation centres, Final report for the Department for Transport, 2 November 2005.

[BRO 00] BEAULIEU M., "Définir et maîtriser la complexité des réseaux de logistique à rebours", *Les 3es Rencontres Internationales de la Recherche en Logistique (RIRL)*, Trois-Rivières, Quebec, 9,10 and 11 May 2000.

[BRU 04] BRUEL O., "Fonction achats: un modèle pour conduire le changement?", *Les Echos, L'Art du Management*, 7 October 2004.

[BRU 07] BRUEL O. (with the participation of ALLEAUME A., AZAMBRE G., BEAUGÉ Th., HUART G., MACADRÉ J., MEDJAD K., PETIT F.), *Management des achats: décisions stratégiques, structurelles et opérationnelles*, Economica, Paris, 2007.

[BRU 05] BRUEL O., PETIT F., "Fonction achats: mesure et pilotage des performances – benchmarking des pratiques", *Cahier de Recherche HEC*, département MIL, v4, 20 January 2005.

[BUR 00] BURLAUD A., "A la recherche d'un système de mesure des performances: application aux réseaux", in FABBE-COSTES N., COLIN J., PACHÉ G. (eds.), *Faire de la recherche en logistique et distribution?*, Vuibert-Fnege, Paris, p. 261-272, 2000.

[CAD 95] CADIOU Y., "Quelles sont la place réelle et les limites d'un tableau de bord logistique dans les décisions stratégiques des entreprises?", *Actes des Premières Rencontres Internationales de la Recherche en Logistique*, Marseille, France, 1995.

[CAL 98] CALVI R., "La gestion des achats en JAT/Flux tendus: approche typologique d'un changement organisationnel", *Gestion 2000*, vol. 14, no. 4, p. 83-105, 1998.

[CAL 99] CALVI R., "L'externalisation des activités d'achat : l'apport du modèle contractualiste appliqué au processus achat", *Finance, Contrôle, Stratégie*, vol. 2, no. 1, p. 27-48, 1999.

[CAL 00A] CALVI R., "Le rôle des Services Achats dans le développement des produits nouveaux : une approche organisationnelle", *Finance, Contrôle, Stratégie*, vol. 3, no. 2, p. 31-55, 2000.

[CAL 00B] CALVI R., LEDAIN M.A., HARBI S., "Le pilotage des partenariats client-fournisseur de l'industrie", *Revue Française de Gestion Industrielle*, vol. 19, no. 1, p. 5-16, 2000.

[CAM 08] CAMMAN C., LIVOLSI L., "Dynamique concurrentielle sur le secteur du transport de marchandises : une approche par les compétences", *7es Rencontres Internationales de Recherche en Logistique*, Avignon, France, 24-26 September 2008.

[CAR 98] CARTER C.R., ELLRAM L.M., "Reverse logistics: a review of the literature and framework for future investigation", *Journal of Business Logistics*, vol. 19, no. 1, p. 85-102, 1998.

[CAR 08] CARTER C.R., ROGERS D.S., "A framework of sustainable supply chain management: moving toward new theory", *International Journal of Physical Distribution & Logistics Management*, vol. 38, no.5, p. 360-387, 2008.

[CIL 08] CILIBERTI F., PONTRANDOLFO P., SCOZZI B., "Investigating corporate social responsibility in supply chain: a SME perspective", *Journal of Cleaner Production*, vol. 16, p. 1579-1588, 2008.

[CHE 04] CHEDOTEL F., "Avoir le sentiment de faire partie d'une équipe : de l'identification à la coopération", *M@n@gement*, vol. 7, no. 3, p. 161-193, 2004.

[CHE 06] CHENG L.C., GRIMM C., "The application of empirical strategic management research to supply chain management", *Journal of Business Logistics*, vol. 27, no. 1, p. 1-55, 2006.

[CHE 98] CHEVALIER A., KUNSCH P.L.,"A multiattractor model for the dynamic control of companies", *Les Cahiers de Recherche de l'ESCP*, no. 98-138, Paris, 1998.

[CHI 94] CHIAPELLO E., DELMOND M.H., "Les tableaux de bord de gestion, outils d'introduction du changement", *Revue Française de Gestion*, no. 97, p. 49-58, 1994.

[CHI 96] CHIAPPELLO E., LEBAS M., "The tableau de bord. A French approach to management information", *19th Annual Meeting of the European Accounting Association*, Bergen, Norway, 2-4 May 1996.

[CHO 06] CHOUINARD M., D'AMOUR S., AÏT-KADI D., Conception et pilotage de boucle d'approvisionnement. Contexte de déploiement d'un réseau d'attribution, de maintenance, de récupération, de traitement et de redistribution de fauteuils roulants, Document de travail DT-2006-SD-01, Laval University, Quebec, 2006.

[CHO 94] CHOW G., HEAVER T., HENRIKSSON L., "Logistics performance: definition and measurement", *International Journal of Physical Distribution and Logistics Management*, vol. 24, no. 1, p. 17-28, 1994.

[CHR 85] CHRISTOPHER M., *The Strategy of Distribution Management*, Gower, Aldershot, 1985.

[CHR 92] CHRISTOPHER M., *Logistics and Supply Chain Management*, Pitman Publishing, London, 1992.

[CHR 99] CHRISTOPHER M., "Les enjeux d'une supply chain globale", *Logistique & Management,* vol. 7, no. 1, p. 3-6, 1999.

[CLA 04] CLAYE-PUAUX S., "Le caractère stratégique de la GRH dans les chaînes logistiques amont tendues", *Les 5es Rencontres Internationales en Recherche Logistique*, Fortaleza, Brazil, 2004.

[COL 96] COLIN J., "La logistique : histoire et perspectives", *Logistique & Management*, vol. 4, no. 2, p. 97-111, 1996.

[COL 02] COLIN J., "De la maîtrise des opérations logistiques au supply chain management", *Gestion 2000*, vol. 19, no. 1, p. 59-74, 2002.

[COL 05] COLIN J., "Le supply chain management existe-il réellement?", *Revue Française de Gestion*, vol 3, no. 156, p. 135-149, 2005.

[COL 88] COLIN J., PACHÉ G., *La logistique de distribution: l'avenir du marketing*, Chotard et Associés Editeurs, Paris, 1988.

[CRO 89] CROSS K., LYNCH R.,"Accounting for competitive performance", *Journal of Cost Management for the Manufacturing Industry*,vol. 3, no. 1, p. 20-28, 1989.

[COO 97] COOPER M., LAMBERT D., PAGH J., "Supply chain management: more than a new name for logistics", *The International Journal of Logistics Management*, vol. 8, no. 1, p. 1-13, 1997.

[DAM 04] DAMAK-AYADI S., "Le référentiel de normalisation SA 8000: enjeux et perspectives", *25e Congrès de l'Association Francophone de Comptabilité*, Orléans, France, 2004.

[DAW 95] DAWE R.L., "Reengineer your returns", *Transportation & Distribution*,vol. 36, no. 8, p. 78, 1995.

[DEP 04] DEPOERS F., "Reporting environnemental: une comparaison GRI/NRE", *Revue Française de Comptabilité*, March, p. 364-365, 2004.

[DEP 10] DEPOERS F., "La consolidation des données environnementales: enjeux et pratiques", *Gestion 2000*, vol. 27, no. 4, p. 129-140, 2010.

[DEP 03] DEPOERS F., REYNAUD E., SCHNEIDER MAUNOURY G., "Comment mesurer la performance durable des entreprises? Proposition d'une grille d'indicateurs", *Gestion 2000*, March-April, p. 13-29, 2003.

[DES 00] DES GARETS V., "Les relations distributeur-fabricant et les modalités de leur coordination", in FABBE-COSTES N., COLIN J., PACHÉ G. (eds.), *Faire de la recherche en logistique et distribution?*, Vuibert-Fnege, Paris, p. 109-126, 2000.

[DON 95] DONALDSON T., PRESTON L., "The stakeholder theory of the corporation: concepts, evidence, and implications", *Academy of Management Review*, vol. 20, no.1, p. 65-91, 1995.

[DUR 10] DURAND B., GONZALES-FELIU J., HENRIOT F., "La logistique urbaine, facteur-clé de développement du B to C", *8e Rencontres Internationales de la recherche en Logistique*, Bordeaux, France, 29-30 September and 1 October 2010.

[EDV 97] EDVINSSON L., MALONE M., *Intellectual Capital, Realizing your Company's True Value by Finding its Hidden Brainpower*, Harper Collins Publishers, New York, NY, 1997.

[ELA 08] ELAMRANI R., CLERGEAU C., BIDAN M., "Le rôle du système d'information dans le pilotage d'un pôle de compétitivité", *Colloque Association Internationale de Management Stratégique*, Nice, France, 2008.

[ELK 10] EL KORCHI A., MILLET D., "Conditions stratégiques d'émergence d'une reverse supply chain fondée sur le remanufacturing chez le fabricant d'origine", *8es Rencontres Internationales de Recherche en Logistique et SCM*, Bordeaux, France, 29-30 September, 1 October 2010.

[ENJ 12] ENJEUX LES ECHOES, http://www.lesechos.fr/enjeux/

[EST 99] ESTAMPE D. (ed.), HARREGUY A., MALDES V., MAMÈRE M., NOUGARET S., TRUIN B.F., "La performance supply chain des acteurs du secteur automobile en France", *Logistique & Management*, vol. 7, no.1, p. 15-22, 1999.

[ETR 07] ETRE C., "Valeur juridique des codes de conduite d'entreprise et étude de leur portée au sein de la distribution", *1e Journée de Recherche Relations entre Industrie et Grande Distribution Alimentaire Comindus*, Avignon, France, 27 March 2007.

[FAB 00] FABBE-COSTES N.,"Supply chain management : concepts et pratiques", *Conférence-débat à l'IAE*, Aix-en-Provence, France, April 2000.

[FAB 07] FABBE-COSTES N., "Systèmes d'information logistique et transport", *Techniques de l'Ingénieur*, 2007.

[FAB 01] FABBE-COSTES N., LEMAIRE C., "La traçabilité totale d'une supply chain: principes, obstacles et perspectives de mise en œuvre", *Revue Française de Gestion Industrielle*, vol. 20, no. 3, p. 23-52, 2001.

[FEN 05] FENNETEAU H., NARO G., "Contrôle de confiance dans l'entreprise virtuelle : illustrations logistiques", *Revue Française de Gestion*, no. 156, p. 203-219, 2005.

[FER 00] FERNANDEZ A., *Les nouveaux tableaux de bord des décideurs*, 2nd edition, Editions d'Organisation, Paris, 2000.

[FRA 06] FRANÇOIS-LECOMPTE A., VALETTE-FLORENCE P., "Mieux connaître le consommateur socialement responsable", *Décisions Marketing*, no. 41, p. 67-79, 2006.

[FUL 11] FULCONIS F., PACHÉ G., ROVEILLO G., *La prestation logistique – Origines, enjeux et perspectives*, Editions EMS management et société, Cormelles-le-Royal, France, 2011.

[GAR 09] GARCIA A., "Identification organisationnelle dans les cabinets d'audit: Quelle place pour la cible opérationnelle?", *30ème congrès de l'Association Francophone de Comptabilité*, Strasbourg, France, 27-29 May 2009.

[GER 05] GERMAIN C., "Une typologie des tableaux de bord implantés dans les petites et moyennes entreprises", *Finance Contrôle Stratégie*, vol. 8, no. 3, p. 125-143, 2005.

[GER 98] GERVAIS M., THENET G., "Planification, gestion budgétaire et turbulence", *Finance Contrôle Stratégie*, vol. 1, no.3, p. 57-84, 1998.

[GHE 05] GHERRA S., "Développement Durable, supply chain management et stratégie: le cas de l'éco-conception", *Logistique & Management*, vol. 13, no. 1, p. 37-48, 2005.

[GIA 03] GIARD V., *Gestion de la production et des flux*, Economica, Paris, 3rd edition, 2003.

[GIB 05] GIBSON B., MENTZER J., COOK R., "Supply Chain Management: the pursuit of a consensus definition", *Journal of Business logistics*, vol. 26, no. 2, p. 17-25, 2005.

[GON 06] GOND J.P., "Gestion des ressources humaines et Développement Durable", in REYNAUD E. (ed.), *Le Développement Durable au cœur de l'entreprise*, Dunod, Paris, p. 83-116, 2006.

[GON 10A] GONZALEZ-FELIU J., MORANA J., "La mutualisation des livraisons en milieu urbain : une réflexion sur les marges de manœuvre du groupe NMPP", *Revue Française de Gestion Industrielle*, numéro spécial sur la logistique urbaine, vol. 29, no. 2, p. 71-91, 2010.

[GON 10B] GONZALEZ-FELIU J., MORANA J., "Are city logistics solutions sustainable? The Cityporto case", *TeMA –Trimestrale del Laboratorio Territorio Mobilità Ambiente*, 03 (02/10), p. 55-64, 2010.

[GRL 95] GRLT – GLOBAL LOGISTICS RESEARCH TEAM, *World Class Logistics: the Challenge of Managing Continuous Change*, Council of Logistics Management, Oak Brook, 1995.

[GUN 01] GUNASEKARAN A., PATEL C., TIRTIROGLU E., "Performance measures and metrics in a supply chain environment", *International Journal of Operations & Production Management*, vol. 21, no. 1-2, p. 71-87, 2001.

[HAR 96] HARLAND C., "Supply chain management: relationships, chains and networks", *British Journal of Management*, vol. 7, Special issue, p. 63-80, 1996.

[HAM 95] HAMEL G., PRAHALAD C.K., *La conquête du futur*, InterEditions, Paris, 1995.

[HAY 79A] HAYES R.H., WHEELWRIGHT S.C., "Link manufacturing process and product life cycles", *Harvard Business Review*, vol. 57, no. 1, p. 133-140, 1979a.

[HAY 79B] HAYES R.H., WHEELWRIGHT S.C., "The dynamics of process-product life cycles", *Harvard Business Review*, vol. 57, no. 2, p. 127-136, 1979b.

[HER 06] HERRBACH O., MIGNONAC K., SIRE B., "Identification ou implication organisationnelle ? Enjeux théoriques et implications pour la recherche en GRH", *Revue de Gestion des Ressources Humaines*, no. 59, p. 3-16, 2006.

[HES 73] HESKETT J., IVIE R., GLASKOWSKY N., *Business logistics*, 2nd edition, The Ronald Press, New York, 1973.

[HIL 94] HILL T., *Manufacturing Strategy, Text and Cases*, 2nd edition, Irwin McGraw-Hill, Boston, MA, 1994.

[HIL 92] HILL C., JONES T., "Stakeholder – Agency Theory", *Journal of Management Studies*,vol. 29, no. 2, p. 131-154, 1992.

[HOC 01] HOCKERS K., "Corporate sustainability management, towards controlling corporate ecological and social sustainability", *Proceedings of Greening of Industry Network Conference*, Bangkok, 21-24 January 2001.

[HOQ 00] HOQUE Z., JAMES W., "Linking balanced scorecard measures to size and market factors: impact on organizational performance", *Journal of Management Accounting Research*, vol. 12, p. 1-17, 2000.

[IGA 01] IGALENS J., JORAS M., *La responsabilité sociale de l'entreprise*, Editions d'Organisation, Paris, 2001.

[JOS 83] JOSSE M., Méthodologie de la modélisation appliquée à la logistique des PMI, Rapport pour le compte de la DGRST, CRET, Aix en Provence, 1983.

[KAP 92] KAPLAN R., NORTON D., "The balanced scorecard – measures that drive performance", *Harvard Business Review*, vol. 70, no. 1, p. 71-79, 1992.

[KAP 96] KAPLAN R., NORTON D., *The Balanced Scorecard*, Harvard Business School Press, Boston, MA, 1996.

[KAP 01] KAPLAN R., NORTON D., *The Strategy-Focused Organization: How Balanced Scorecard Companies Thrive in the New Business Environment*, Harvard Business School Press, Boston, MA, 2001.

[KAP 06] KAPLAN R., NORTON D., *Alignment – Using the Balanced Scorecard to Create Corporate Synergies*, Harvard Business School Press, Boston, MA, 2006.

[KAR 10] KARÂA M., Les déterminants de l'adoption de la traçabilité par les entreprises de conditionnement de dattes en Tunisie, Thesis, Management Sciences, University of Aix-Marseille 2, 2 July 2010.

[LAL 96] LALONDE B., POHLEN T., "Issues in supply chain costing", *The International Journal of Logistics Management*, vol. 7, no. 1, p. 1-12, 1996.

[LAM 94] LAMBERT D., "Pour une stratégie logistique orientée client", *Logistique & Management*, vol. 2, no. 1, p. 59-71, 1994.

[LAM 98] LAMBERT D., COOPER M., PAGH J., "Supply chain management: implementation issues and research opportunities", *The International Journal of Logistics Management*, vol. 9, no.2, p. 1-19, 1998.

[LAM 05] LAMBERT D., GARCIA-DASTUGUE S., CROXTON K., "An evaluation of process-oriented supply chain management frameworks", *Journal of Business Logistics*, vol. 26, no.1, p. 25-51, 2005.

[LAM 03] LAMBERT S., RIOPEL D., "Logistique inverse : revue de littérature", *Les Cahiers du GERAD*, no. G-2003-61, October 2003.

[LAM 04] LAMBERT S., RIOPEL D., "Modèle intégrateur de la logistique inverse", *Les Cahiers du GERAD*, no. G-2004-69, August 2004.

[LEB 95] LEBOTERF G., *De la compétence*, Editions d'Organisation, Paris, 1995.

[LEM 02] LEMOINE C., *Se former au bilan de compétences*, Dunod, Paris, 2002.

[LIB 98] LIBERATONE M.J., MILLER T., "A framework for integrating activity-based costing and the balanced scorecard into the logistics strategy development and monitoring process", *Journal of Business Logistics*, vol. 19, no. 2, p. 131-154, 1998.

[LIV 06] LIVOLSI L., CAMMAN C., ROUSSAT C., *Lexipro. Le lexique des termes de la logistique*, 5th edition, Groupe Liaisons, Paris, 2006.

[LOA 97] LOUART P., "Motivation", in SIMON Y., JOFFRE P. (eds), *Encyclopédie de gestion*, p. 2095-2107, 2nd edition, Economica, Paris, 1997.

[LOG 10] LOGISTIQUES MAGAZINE, http://www.wk-transport-logistique.fr/consultation/sommaire.php?acronyme=lm

[LOG 11] LOGISTIQUES MAGAZINE, http://www.wk-transport-logistique.fr/consultation/sommaire.php?acronyme=lm

[MAC 05] MACDONALD J.P., "Strategic sustainable development using the ISO14001 standard", *Journal of Cleaner Production*, vol. 13, no. 6, p. 631-643, 2005.

[MAL 03] MALTZ A., SHENHAR A., REILLY R., "Beyond the balanced scorecard: refining the search for organizational success measures", *Long Range Planning*, vol. 36, p. 187-204, 2003.

[MEN 01] MENTZER J., DEWITT W., KEEBLER J., MIN S., NIX N., SMITH C., ZACHARIA Z., "Defining supply chain management", *Journal of Business Logistics*, vol. 22, no. 2, p. 1-25, 2001.

[MER 02] MERCIER S., "Une typologie de la formalisation de l'éthique en entreprise: l'analyse de contenu de 50 documents", *Revue de gestion des ressources humaines*, no. 43, p. 34-49, 2002.

[MES 99] MESNARD X., DUPONT A., "Votre logistique est-elle à la pointe?", *L'Expansion Management Review*, no. 94, p. 52-58, 1999.

[MIC 10] MICHAUD C., Comportements des consommateurs et biens privés à caractéristiques environnementales: une approche expérimentale, Thesis, Economic Sciences, University of Grenoble, 1 March 2010.

[MIC 05] MICHALSKA J., "The usage of the balanced scorecard for the estimation of the enterprise's effectiveness", *Journal of Materials Processing Technology*, vol. 162-163, p. 751-758, 2005.

[MIL 91] MILAN G., "La plate-forme stratégique dans un environnement instable", *Revue Française de Gestion*, no. 86, p. 57-60, 1991.

[MIL 03] MILLET D. (ed.), COPPENS C., JACQUESON L., LE BORGNE R., TONNELIER P., *Intégration de l'environnement en conception*, Hermès, Paris, 2003.

[MIL 95] MILTENBURG J., *Manufacturing Strategy: How to Formulate and Implement a Winning Plan*, Productivity Press, Portland, 1995.

[MIN 04] MIN S., MENTZER J., "Developing and measuring supply chain management concepts", *Journal of Business Logistics*, vol. 25, no.1, p. 63-91, 2004.

[MIN 99] MINTZBERG H., AHLSTRAND B., LAMPEL J., *Safari en pays stratégie*, Village Mondial, Paris, 1999.

[MIY 99] MIYASAKI A., PHILIPPS J., PHILIPPS D., "Twenty years of JBL: an analysis of published research", *Journal of Business logistics*, vol. 20, no.2, p. 1-19, 1999.

[MOC 06] MOCELLIN F., Gestion des entrepôts et plates-formes – Assurer la performance de votre supply chain par la maîtrise des zones de stockage, 2nd edition, Dunod, Paris, 2006.

[MON 94] DE MONTMOLLIN M., L'intelligence de la tâche. Eléments d'ergonomie cognitive, Peter Lang, Bern, 1994.

[MON 97] DE MONTMORILLON B., "Gestion de l'emploi dans l'organisation productive", *Encyclopédie de gestion*, t. 2, p. 1482-1499, 2nd edition, Economica, Paris, 1997.

[MOR 02] MORANA J., Le couplage supply chain management – tableau de bord stratégique : une approche exploratoire, Doctoral Thesis, University of the Mediterranean Aix-Marseille II, 2 October 2002.

[MOR 04A] MORANA J., "10 ans de Logistique et Management : une analyse des publications de recherche", *Logistique & Management*, vol. 12, no. 1, p. 3-21, 2004.

[MOR 04B] MORANA J., "Le capital humain dans le réseau d'affaires: une réflexion dans la pratique d'un EDI (Electronic Data Interchange) au sein d'une PME", *Revue Humanisme & Entreprise*, no.267, p. 89-108, 2004.

[MOR 05] MORANA J., "Quels auteurs citons-nous ?", *Logistique & Management*, vol. 12, no.2, p. 65-68, 2005.

[MOR 07] MORANA J., "La démarche Supply Chain Management vue par les salariés d'une multinationale de la haute technologie", *Logistique & Management*, vol. 15, no.2, p. 81-93, 2007.

[MOR 08] MORANA J., "L'utilisation d'indicateurs logistiques : une étude exploratoire via le modèle SCOR® (Supply-Chain Operations Reference Model®)", *Logistique & Management*, vol. 16, no.2, p. 31-44, 2008.

[MOR 10] MORANA J., "Les 16 ans de la revue Logistique & Management", *8es Rencontres Internationales de Recherche en Logistique (RIRL)*, Bordeaux, France, 29-30 September, 1 October 2010.

[MOR 11] MORANA J., GONZALEZ-FELIU J., "Interporto di Padova S.P.A.: un camion vert dans la ville", in JOFFRE O., SIMON E. (eds.), *Revue de cas en gestion*, EMS Management & Société, Cormelles-le-Royal, 2011.

[MOR 00] MORANA J., PACHÉ G., "Supply chain management et tableau de bord prospectif: à la recherche de synergies", *Logistique & Management*, vol. 8, no. 1, p. 77-88, 2000.

[MOR 08] MORANA J., VAN HOOREBEKE D., PIRE-LECHALARD P., "Supply chain management, green SCM, social SCM: la logistique globale au cœur au développement durable", *3es journées Neptune et 1e université "Réalités et Prospectives du Développement Durable organisationnel"*, Toulon, France, 13-14 November 2008.

[MOU 12] MOULETTE P., ROQUES O., *Maxi fiches de gestion des ressources humaines*, Dunod, Paris, 2012.

[MUL 04] MULLER J., "Les déterminants de la performance de la fonction achat industriel: définitions, tests et implications", *La Revue des Sciences de Gestion, Direction et Gestion*, no. 208-209, p. 111-124, 2004.

[MUR 03] MURPHY P., POIST R., "Green perspectives and practices: a "comparative logistics" study", *Supply Chain Management: an International Journal*, vol. 8, no. 2, p. 122-131, 2003.

[NEW 10] New Press, http//www.newpress.fr/

[NOR 00] NORREKLIT H., "The balance on the balanced scorecard – a critical analysis of some of its assumptions", *Management Accounting Research*, vol. 11, no.1, p. 65-88, 2000.

[NOR 03] NORREKLIT H., "The balanced scorecard: what is the score? A rhetorical analysis of the balanced scorecard accounting", *Organizations and Society*, vol. 28, p. 591-619, 2003.

[OHN 89] OHNO T., *L'esprit Toyota*, Masson, Paris, 1989.

[ORS 03] ORSE (OBSERVATOIRE SUR LA RESPONSABILITÉ SOCIÉTALE DES ENTREPRISES), Analyse comparative d'indicateurs du développement durable, available at www.industrie.gouv.fr/pdf/devdurable1.pdf, 2003.

[PAC 08] PACHÉ G., "Efficient urban e-logistics: mutualization of resources and source of competitive advantage", *7th International Meeting for Research in Logistics*, Avignon, France, 24-26 September 2008.

[PAC 99] PACHÉ G., COLIN J., "Exploring the evolution of the logistics channel in electronic commerce: a research program", *Proceedings of the Ninth Biennial World Marketing Congress*, p. 5-10, Qawra, Malta, June 1999.

[PAC 00] PACHÉ G., COLIN J., "Recherche et applications en logistique : des questions d'hier, d'aujourd'hui et de demain", in FABBE-COSTES N., COLIN J., PACHÉ G. (eds), *Faire de la recherche en logistique et distribution?*, p. 31-53, Vuibert-Fnege, Paris, 2000.

[PAC 97] PACHÉ G., DESGARETS V., "Relations inter-organisationnelles dans les canaux de distribution: les dimensions logistiques", *Recherche et Applications en Marketing*, vol. 12, no. 2, p. 61-82, 1997.

[PAC 06] PACHÉ G., PARAPONARIS C., *L'entreprise en réseau: approches intra et inter-organisationnelles*, Edition de l'Adreg, Paris, 2006.

[PAL 00] PALMERO S., "Implication organisationnelle, satisfaction au travail, engagement au travail et intention de départ des salariés à temps partiel", *Actes du 11e congrès de l'AGRH*, Paris, 16-17 November 2000.

[PAT 09] PATIER D., ROUTHIER J.L., "Une méthode d'enquête du transport de marchandises en ville pour un diagnostic en politiques urbaines", *Les Cahiers Scientifiques du Transport*, no. 55, p. 11-38, 2009.

[PEL 07] PELLATON A., VIRUEGA J.L., "L'utilisation de la traçabilité pour la sécurité sanitaire: analyse par la théorie de la traduction", *1e journée de recherche Relations entre Industrie et Grande Distribution Alimentaire*, Avignon, France, 29 March 2007.

[PER 98] PERSONNE M., Contribution à la méthodologie d'intégration de l'environnement dans les PME-PMI: évaluation des performances environnementales, Doctoral Thesis, INSA, Lyon, 1998.

[PET 06] PETITQUEUX A., "Implémentation Lean : application industrielle", *Les Techniques de l'Ingénieur*, 2006.

[PIE 73] PIÉRON H., *Vocabulaire de la psychologie*, PUF, Paris, 1973.

[PIT 99] PITTIGLIO T., RABIN R., TODD R., MCGRATH M., "Supply chain: mode d'emploi. Les bonnes pratiques du supply chain management", *Logistiques Magazine*, dossier, June 1999.

[PON 93] PONS J., CHEVALIER P., *La logistique intégrée*, Hermès, Paris, 1993.

[POR 82] PORTER M., *Choix stratégiques et concurrence*, Economica, Paris, 1982.

[POR 85] PORTER M., *Competitive Advantage*, The Free Press, New York, 1985.

[POT 98] POTAGE J., "Les achats à Thomson-CSF, vers un nécessaire modèle de maturité", *Revue Internationale de l'Achat*, vol. 18, no. 2, p. 1118, 1998.

[PRA 90] PRAHALAD C.K., HAMEL G., "The core competence of the corporation", *Harvard Business Review*, vol. 68, no. 3, p. 79-91, 1990.

[PRA 98] PRATT M.G., "To be or not to be: central questions in organizational identification", in WHETTEN A.D., GODFREY P.C. (eds.), *Identity in Organizations. Building Theory Through Conversations*, p. 171-207, Sage Publications, Thousand Oaks, London, New Delhi, 1998.

[PRO 95] PROBST G., BÜCHEL B., *La pratique de l'entreprise apprenante*, Editions d'Organisation, Paris, 1995.

[REI 95] REIX R., *Systèmes d'information et management des organisations*, Vuibert, Paris, 1995.

[REI 97] REIX R., "Flexibilité", in JOFFRE P., SIMON Y. (eds), *Encyclopédie de gestion*, p. 1407-1420, 2nd edition, Economica, Paris, 1997.

[REY 07] REYES CARRILLO T., L'éco-conception dans les PME: Les mécanismes du cheval de Troie méthodologique et du choix de trajectoires comme vecteurs d'intégration de l'environnement en conception, Thesis, Industrial Engineering, University of Sud Toulon-Var, 14 December 2007.

[RIN 89] RINEHART L., COOPER B., WAGENHEIM G., "Furthering the integration of marketing and logistics through customer service in the channels", *Journal of the Academy of Marketing Science*, vol. 17, no.1, p. 63-71, 1989.

[ROM 00] ROMEYER C., "Le rôle de la traçabilité de l'activité dans la gestion de la chaîne logistique intra-organisationnelle", *Les 3es Rencontres Internationales de la Recherche en Logistique*, Trois-Rivières, Quebec, 9-11 May 2000.

[ROG 98] ROGERS D., TIBBEN-LEMBKE R., "Going backwards: reverse logistics trends and practices", *Reverse Logistics Executive Council*, 1998.

[ROG 99] ROGERS D., TIBBEN-LEMBKE R., "Reverse logistics: stratégies et techniques", *Logistique & Management*, vol. 7, no.2, p. 15-25, 1999.

[ROG 01] ROGERS D., TIBBEN-LEMBKE R., "An examination of reverse logistics practices", *Journal of Business Logistics*, vol. 22, no.2, p. 129-147, 2001.

[ROL 99] ROLLE J.D., PILLET J.L., "La mesure du potentiel EDI en entreprise. Le cas des fournisseurs du CERN", *Système d'Information et Management*, vol. 4, no. 2, p. 65-90, 1999.

[SAU 00] SAUVAGE T., NAHON D., "Les politiques fournisseurs dans l'automobile et la grande distribution: réflexions sur le pouvoir et la dépendance", in FABBE-COSTES N., COLIN J., PACHÉ G. (eds.), *Faire de la recherche en logistique et distribution?*, p. 159-169, Vuibert-Fnege, Paris, 2000.

[SCO 91] SCOTT C., WESTBROOK R., "New strategic tools for supply chain management", *International Journal of Physical Distribution and Logistics Management*, vol. 21, no.1, p. 23-33, 1991.

[SEN 90] SENGE P., *La cinquième discipline*, First, Paris, 1990.

[SEN 02] SENKEL M.P., "Coordination logistique et fonctionnement du canal de distribution", *Revue Française de Marketing*, no.188, p. 79-89, 2002.

[SEN 08] SENKEL M.P., KOLEVA P., "Les responsables logistiques face à la responsabilité sociale de l'entreprise : premiers résultats d'une recherche exploratoire", *7es Rencontres Internationales de la Recherche en Logistique*, Avignon, France, 24-26 September 2008.

[SEU 08] SEURING S., MÜLLER M., "From a literature review to a conceptual framework for sustainable supply chain management", *Journal of Cleaner Production*, vol. 16, no. 15, p. 1699-1710, 2008.

[SIM 03] SIMONET E., "Les systèmes de management environnemental", *Synthèse*, available at www.laplateforme.org/IMG/pdf/SME_synthese.pdf, 2003.

[SPE 03] SPECKBACHER G., BISCHOF J., PFEIFFER T., "A descriptive analysis on the implementation of balanced scorecards in German-speaking countries", *Management Accounting Research*, vol. 14, p. 361-387, 2003.

[STA 92] STALK G., EVANS P., SHULMAN L., "Competing on capabilities: the new rules of corporate strategy", *Harvard Business Review*, vol. 70, no. 2, p. 57-69, 1992.

[STE 96] STEWART G., "Optimiser vos coûts logistiques grâce à une modélisation simple et efficace", *Logistique & Management*, vol. 4, no. 2, p. 65-72, 1996.

[SUP 02] SUPIZET J., "Total balanced scorecard, un pilotage aux instruments", *L'Informatique Professionnelle*, no. 209, p. 15-20, 2002.

[SRI 07] SRIVASTAVA S., "Green supply-chain management: a state-of-the-art literature review", *International Journal of Management Review*, vol. 9, no. 1, p. 53-80, 2007.

[TAR 99] TARONDEAU J.C., *La flexibilité dans les entreprises*, Presses Universitaires de France, Paris, 1999.

[TCH 95] TCHOKOGUÉ A., "Fonction logistique et management financier de l'entreprise", *Logistique & Management*, vol. 3, no.1, p. 45-58, 1995.

[TIX 83] TIXIER D., MATHE H., COLIN, J., *La logistique au service de l'entreprise*, Dunod, Paris, 1983.

[TRA 12] Transport and Logisitique, http://www.wk-transport-logistique.fr/

[ULR 91] ULRICH D., LAKE D., "Organizational capabilities: creating competitive advantage", *The Executive*, vol. 5, no. 1, p. 77-93, 1991.

[USI 12] L'Usine Nouvelle, http://usinenouvelle.com/

[VAN 08] VAN HOOREBEKE D., PIRÉ-LECHALARD P., "La socio-conception et sa gouvernance", *Colloque Institut d'Audit Social, "Audit social et gouvernance"*, Tangiers, Morocco, 8-10 May 2008.

[VAN 96] VAN WEELE A., *Purchasing and Supply Chain Management*, International Thomson Business Press, London, 1996.

[VAN 98] VANDAELE M., "L'ECR : moyen indispensable du partenariat producteurs-distributeurs", *Décisions Marketing*, no.15, p. 69-76, 1998.

[VIN 98] VINCENT CL.., "L'e-commerce, c'est dépassé", *Enjeux Les Echos*, June, p. 32-41, 2012.

[VIR 04] VIRUEGA J.L., *Traçabilité: Outils, méthodes et pratiques*, Editions d'Organisation, Paris, 2004.

[ZUT 04] ZUTSHI A., SOHAL A., "A study of the environmental management system (EMS) adoption process within Australian organisations – 2. Role of Stakeholders", *Technovation*, vol. 24, no. 5, p. 371-386, 2004.

Acronyms and Abbreviations

1PL	First Party Logistics (Provider)
2PL	Second Party Logistics (Provider)
3E	Electrical and Electronic Equipment
3PL	Third Party Logistics (Provider)
4PL	Fourth Party Logistics (Provider)
5PL	Fifth Party Logistics (Provider)
2SoSCM	Social/Societal Supply Chain Management
ADEME	*Agence de l'Environnement et de la Maîtrise de l'Energie* French Agency for the Environment and Control of Energy)
AFNOR	*Association Française de Normalisation* (French Standardization Association)
APS	Advanced Planning System
AUTF	*Association des Utilisateurs de Transport de Fret* (French Association for Users of Freight Transport)
BLU	*Boîte Logistique Urbaine* (urban logistics box)
BSC	Balanced Scorecard
CDU	*Centre de Distribution Urbain* (urban distribution center)
CJDES	*Centre des Jeunes Dirigeants et des acteurs de l'Economie Sociale* (Center for Young Managers and Actors in Social Economy)

CMDU	*Centre Multimodal de Distribution Urbaine* (multimodal urban distribution center)
CPFR	Collaborative Planning, Forecasting and Replenishment
CRM	Customer Relationship Management
CSR	Corporate Social Responsibility
DRP	Distribution Replenishment Planning
CSCMP	Council of Supply Chain Management Professionals
ECD	Environmentally Conscious Design
ELV	End-of-Life Vehicle
EEC	European Economic Community
ECR	Efficient Consumer Response
EDI	Electronic Data Interchange
EEE	Electrical and Electronic Equipment
ELP	*Espace Logistique de Proximité* or *Espace de Livraison à Proximité* (local logistics space)
ELU	*Espace Logistique Urbain* (urban logistics space)
EMAS	Eco Management and Audit Scheme
EMS	Environmental Management System
ERP	Enterprise Resource Planning
ERP	European Recycling Platform
FNTR	*Fédération Nationale des Transports Routiers* (French National Road Transport Federation)
FSC	Forest Stewardship Council Scheme
GHG	Greenhouse Gas
GMA	*Gestion Mutualisée des Approvisionnements* (mutualized supply management)
GPA	*Gestion Partagée des Approvisionnements* (shared supply management)
GPEC	*Gestion Prévisionnelle des Emplois et des Compétences* (project management of jobs and skills)

GRI	Global Reporting Initiative
GrSCM	Green Supply Chain Management
HACCP	Hazard Analysis Critical Control Points
HQE®	*Haute Qualité Environnementale* (high environmental quality)
IAS	International Accounting Standards
IFAC	International Federation of ACcountants
IFRS	International Financial Reporting Standards
ILO	International Labor Organization
IS	Information System
ISO	International Standardization Organization
IUCN	International Union for Conservation of Nature
LCA	Lifecycle Assessment/Analysis
LLP	Lead Logistics Provider
MEDDTL	*Ministère de l'Ecologie, du Développement Durable, des Transports et du Logement* (French Ministry of Ecology, Sustainable Development, Transport and Housing)
MES	Manufacturing Execution System
MRP	Material Requirement Planning
MRP2	Manufacturing Resources Planning
NF	*Norme Française* (French Standard)
NHIW	Non-Hazardous Industrial Waste
NRE	*Nouvelles Régulations Economiques* (law) (new economic regulations)
OHSAS	Occupational Health and Safety Assessment Series
OTRE	*Organisation des TPE et PME du transport routier* (French organization for SMEs in the road transport field)
PAV	*Points d'Accueil de Véhicules* (vehicle stopping points or loading bays)
PEFC	Pan-European Forest Certification Scheme

RFID	Radio Frequency Identification
RoPo	Research Online, Purchase Offline
RONA	Return On Net Assets
SAI	Social Accountability International
SCC	Supply Chain Council
SCE	Supply Chain Execution
SCM	Supply Chain Management
SCO	Supply Chain Orientation
SCOR®	Supply Chain Operations Reference-model®
SD	Sustainable Development
SME/SMI	Small and Medium Enterprises/Small and Medium Industry
SMED	Single Minute Exchange of Die
SIW	Special Industrial Waste
SRC	Socially Responsible Consumption
SRI	Socially Responsible Investing
SRM	Supplier Relationship Management
SuSCM	Sustainable Supply Chain Management
SuSCMBSC	Sustainable Supply Chain Management Balanced SCorecard
TLF	*(Union des entreprises de) Transport et de Logistique de France* (French Transport and Logistics Union)
TMS	Transport Management System
TPS	Toyota Production System
TWDQ	Toxic Waste in Dispersed Quantities
UN	United Nations
UNEP	United Nations Environment Programme

UNOSTRA	*Union Nationale des Organisations Syndicales des Transporteurs Routiers Automobiles* (French National Union for Automobile and Road Transport Organizations)
VMI	Vendor Managed Inventory
VSM	Value Stream Mapping
WBCSD	World Business Council for Sustainable Development
WEEE	Waste Electrical and Electronic Equipment
WF	Work Force
WMS	Warehouse Management System
WRI	World Resources Institute
ZLU	*Zones Logistiques Urbaines* (urban logistics zones)

Index